A Milliliter-Scale Setup for the Efficient Characterization of Multicomponent Vapor-Liquid Equilibria Using Raman Spectroscopy

Eine Apparatur auf Milliliter-Skala zur effizienten Charakterisierung multinärer Dampf-Flüssig-Gleichgewichte mittels Raman-Spektroskopie

Von der Fakultät für Maschinenwesen der
Rheinisch-Westfälischen Technischen Hochschule Aachen
zur Erlangung des akademischen Grades eines Doktors
der Ingenieurwissenschaften genehmigte Dissertation

vorgelegt von

Bastian Liebergesell

Berichter: Univ.-Prof. Dr.-Ing. André Bardow
Univ.-Prof. Dr.-Ing. habil. Roland Span

Tag der mündlichen Prüfung: 05. Oktober 2018

Aachener Beiträge zur Technischen Thermodynamik Band 16

Bastian Liebergesell
A Milliliter-Scale Setup for the Efficient Characterization of Multicomponent Vapor-Liquid Equilibria Using Raman Spectroscopy
Eine Apparatur auf Milliliter-Skala zur effizienten Charakterisierung multinärer Dampf-Flüssig-Gleichgewichte mittels Raman-Spektroskopie

ISBN: 978-3-95886-247-0

Bibliografische Information der Deutschen Bibliothek
Die Deutsche Bibliothek verzeichnet diese Publikation in der Deutschen Nationalbibliografie; detaillierte bibliografische Daten sind im Internet über http://dnb.ddb.de abrufbar.

Herstellung & Vertrieb:

1. Auflage 2018

Süsterfeldstr. 83, 52072 Aachen
Tel. 0241/87 34 34
Fax 0241/87 55 77
www.Verlag-Mainz.de

ISSN: 2198-4832

Satz: nach Druckvorlage des Autors
Umschlaggestaltung: Druckerei Mainz

printed in Germany
D82 (Diss. RWTH Aachen University, 2018)

'In God we trust. All others must bring data.'

W. Edwards Deming

Vorwort

Die vorliegende Arbeit entstand im Rahmen meiner Tätigkeit als wissenschaftlicher Mitarbeiter am Lehrstuhl für Technische Thermodynamik der RWTH Aachen. Mein Dank gebührt an erster Stelle André Bardow für die Möglichkeit diese Arbeit an seinem Lehrstuhl anzufertigen. Ich bedanke mich herzlich für das mir entgegengebrachte Vertrauen und die mir übertragene Verantwortung. Beides hat mich persönlich und fachlich entscheidend entwickelt. Des Weiteren gilt mein Dank Prof. Roland Span für die Übernahme des Korreferats im Promotionsverfahren. Ich danke Prof. Ronald Gebhardt für die Übernahme des Prüfungsvorsitzes. Dem Exzellenzcluster *Tailor-Made Fuels from Biomass* danke ich für die Finanzierung des Projektes.
Ich hatte eine großartige Zeit, was zum größten Teil an tollen Kollegen gelegen hat, die täglich wieder dafür sorgen, dass man gerne zur Arbeit kommt! Dafür danke ich allen Mitarbeitern des Lehrstuhls und insbesondere der OMT. Von der alltäglichen Arbeit bis zum Weihnachtsfeierbeitrag hat es mir stets Spaß gemacht. Ich bedanke mich herzlich bei unseren TaMis, ohne die die praktische Umsetzung meiner Arbeit nie möglich gewesen wäre! Einen möchte ich an dieser Stelle hervorheben: Willi, du bist ne Wucht! Danke für deine Unterstützung. Mein großer Dank gilt Sebastian für seine Hilfe und Geduld bei der Modellierung von Phasengleichgewichten. Und natürlich für die Chilis und die Tomaten! Vielen Dank an Paule und Hans-Jürgen für die lehrreichen Stunden, die ich mit euch verbringen durfte. Meinen Kollegen (und noch so viel mehr) Toddy, Thomas und Carsten danke ich für die stets offenen Ohren. Ob sinnvoll oder spaßig, ihr habt mir meine Zeit versüßt. Carsten danke ich besonders für seinen Beitrag an dieser Arbeit: ich kann deinen Anteil nicht hoch genug wertschätzen.
Meiner Mutter gebührt mein größter Dank: für die Freiheiten und die grenzenlose Unterstützung, ohne die ich nicht an diesem Punkt angekommen wäre! Das hier ist für dich!
Maike, ich danke dir für dein Verständnis und deine Unterstützung in den letzten Jahren. Du machst mich sehr glücklich!

Aachen, Oktober 2018 *Bastian Liebergesell*

Contents

List of Figures

List of Tables

Notation

Abbreviations

2-MF	2-methylfuran
2-MTHF	2-methyltetrahydrofuran
a.u.	Arbitrary unit
BHT	Butylated hydroxytoluene
cal	Calculated
CCD	Charge-coupled device
COSMO-RS	Conductor-like Screening Model for Real Solvents
DDB	Dortmund Data Bank
DNBE	di-*n*-butyl ether
EOS	Equation of state
EtOH	Ethanol
exp	Experimental
FD-IHM	First derivative indirect hard modeling
HCCI	Homogeneous charge compression ignition
IHM	Indirect hard modeling
iso-octane	2,2,4-trimethylpentane
liq	Liquid
min	Minutes
MTBE	Methyl tert-butyl ether
N_2	Nitrogen
NOx	Nitrogen oxides
NRTL	non-random two-liquid model
PCP-SAFT	Perturbed-Chain Polar Statistical Associating Fluid Theory
PR	Peng-Robinson
RAMSPEQU	Raman Spectroscopic Phase Equilibrium Characterization
RMSECV	Root mean squared error of cross-validation
SRK	Soave-Redlich-Kwong
THF	Tetrahydrofuran
TMFB	Tailor-Made Fuels from Biomass

UNIFAC	UNIQUAC Functional-group Activity Coefficients
UNIQUAC	Universal quasi-chemical model
vap	Vapor
vdW	van-der-Waals
VLE	Vapor-liquid equilibrium
VLLE	Vapor-liquid-liquid equilibrium

Symbols

A	Helmholtz free energy	J
A^{chain}	Helmholtz free energy contribution of chain group interactions	J
A^{disp}	Helmholtz free energy contribution of dispersion interactions	J
A^{DM}	Helmholtz free energy contribution of dipolar interactions	J
A^{hs}	Helmholtz free energy contribution of hard sphere interactions	J
A^{QM}	Helmholtz free energy contribution of quadrupolar interactions	J
A^{res}	Residual Helmholtz free energy	J
f	fugacity	Pa
g	standard gravity	$\text{m}\,\text{s}^{-2}$
G	Gibbs free energy	J
G^{E}	excess Gibbs energy	J
k	Calibration constant	
k_{B}	Boltzmann constant	$\text{J}\,\text{K}^{-1}$
$k_{i,j,\text{phase}}$	Calibration constant for components i and j in the indicated phase	
k_{ij}	Binary interaction parameter for components i and j	
m_i	Number of molecular segments of component i	
N	Number of experimentally determined property values	
n_i	Amount of substance of component i	mol
p	Pressure	kPa
$p_{\text{s0},i}(T)$	Vapor pressure of component i (as a function of the temperature T)	kPa
S_i	Raman signal of component i	
T	Temperature	K
$u(i)$	Uncertainty of value i	
v	Molar volume	$\text{m}^3\,\text{mol}^{-1}$
V	Volume	m^3
x	Mole fraction in the liquid phase	
y	Mole fraction in the vapor phase	

Greek symbols

γ_i	activity coefficient of component i in the liquid phase	
ϵ_i	Dispersion parameter of component i	J
ϵ_{ij}	One-fluid mixture dispersion parameter of components i and j	J
Θ_i	Quadrupole moment of component i	DÅ
μ_i^α	Chemical potential of component i in phase α	$\mathrm{J\,mol^{-1}}$
μ_i	Dipole moment of component i	D
ρ	Density	$\mathrm{g\,cm^{-3}}$
σ_i	Segment diameter of component i	Å
φ_i	fugacity coefficient of component i	

Kurzfassung

Dampf-Flüssig-Gleichgewichtsdaten sind von entscheidender Bedeutung für die chemische Industrie. Die Bestimmung von Dampf-Flüssig-Gleichgewichten (VLE) ist typischerweise mit hohem experimentellen Aufwand verbunden. Um diesen Aufwand zu begrenzen, werden VLE häufig durch synthetische Methoden bestimmt. Synthetische Methoden nutzen Proben bekannter Zusammensetzung um Probenentnahme und komplexe Analytik zu vermeiden. Analytische Methoden liefern hingegen Zusammensetzungen aus unabhängigen Messungen. Für diese Messungen werden üblicherweise Proben entnommen. Daher werden für analytische Methoden häufig große Substanzmengen eingesetzt.

Im ersten Teil dieser Arbeit nutzen wir eine etablierte synthetische Methode (Cailletet-Methode) zur Charakterisierung des Hochdruck-VLEs zweier vielversprechender binärer Biokraftstoffgemische. Die Cailletet-Methode entspricht dem Stand der Technik und dient als Referenzmethode. Die Methode ermöglicht Messungen von außergewöhnlicher Genauigkeit, benötigt jedoch eine umfangreiche Infrastruktur.

Um diese umfangreiche Infrastruktur zu vermeiden und Einschränkungen bisheriger VLE-Methoden zu überwinden, wird im zweiten Teil der Arbeit eine neuartige Apparatur zur nicht-invasiven und effizienten Charakterisierung von VLE auf Milliliterskala entwickelt: RAMSPEQU (**Ram**an **S**pectroscopic **P**hase **Equ**ilibrium Characterization). RAMSPEQU ermöglicht eine substanzsparende und schnelle Bestimmung von VLE. Phasenzusammensetzungen werden mittels Raman-Spektroskopie bestimmt, wodurch Probenentnahme und damit verbundene Fehler vermieden werden. Infolgedessen genügen bereits Volumina von unter 3 ml für verlässliche VLE-Messungen. Um substanzsparend eine schnelle Datenerfassung zu ermöglichen, stellen wir einen integrierten Arbeitsablauf vor. Der Arbeitsablauf kombiniert die Kalibration der Raman-Spektroskopie und die eigentliche VLE-Messungen. RAMSPEQU ermöglicht so die Erfassung von bis zu 15 $pTxy$-Datensätzen innerhalb eines Arbeitstages. RAMSPEQU wird durch Vergleich mit Reinstoffdaten und binären VLE aus der Literatur erfolgreich validiert.

Gemische mit nur zwei Komponenten sind für industrielle Anwendungen selten von Interesse. Mit wachsender Anzahl von Komponenten steigt der experimentelle Aufwand für die Messung von VLE überproportional. Daher ist die effiziente RAMSPEQU-Methode für Multikomponentensysteme besonders geeignet. Im dritten Teil der Arbeit wird RAMSPEQU für die Messung des VLE eines quaternären Systems und seiner binären Subsysteme eingesetzt. RAMSPEQU ermöglicht die experimentelle Bestimmung des VLE mit geringen Substanzmengen: 22 ml (binär) beziehungsweise 105 ml (quaternär) sind für eine umfangreiche Beschreibung des multinären VLE ausreichend.

Die RAMSPEQU-Apparatur und ihr integrierter Arbeitsablauf ermöglichen die Charakterisierung von multinären VLE mit signifikanten Einsparungen von Substanz und Arbeitszeit.

Abstract

Vapor-liquid equilibrium (VLE) data are of major importance for the chemical industry. Despite significant progress in predictive methods, experimental VLE data are still indispensable. In this work, we address the need for experimental VLE data. Commonly, the characterization of VLE requires significant experimental effort. To limit the experimental effort, VLE measurements are frequently conducted by synthetic methods which employ samples of known composition and avoid complex analytics and sampling issues. In contrast, analytical methods provide independent information on phase compositions, commonly based on sampling and large amounts of substance.

In the first part of this work, we employ a synthetic method, the well-established Cailletet setup, to characterize the high pressure VLE of two promising binary biofuel blends. The Cailletet method serves as a state of the art reference method that enables collecting data of remarkable accuracy. However, extensive infrastructure is needed.

In the second part, to avoid extensive infrastructure and overcome limitations of previous methods, we develop a novel analytical milliliter-scale setup for the non-invasive and efficient characterization of VLE: RAMSPEQU (**Ram**an **S**pectroscopic **P**hase **Equ**ilibrium Characterization). The novel setup saves substance and rapidly characterizes VLE. Sampling and its associated errors are avoided by analyzing phase compositions using Raman spectroscopy. Thereby, volumes of less than 3 ml are sufficient for reliable phase equilibrium measurements. To enable rapid data generation and save substance, we design an integrated workflow combining Raman signal calibration and VLE measurement. As a result, RAMSPEQU gives access to up to 15 $pTxy$-data sets per workday. RAMSPEQU is successfully validated against pure component and binary VLE data from literature.

However, mixtures with only two components rarely depict real industrial applications. As the number of experiments increases strongly with a rising number of components, the efficient RAMSPEQU setup seems particularly suited for multicomponent systems. In the third part of this work, we employ the RAMSPEQU setup for the characterization of a quaternary system and its binary subsystems. 22 ml and 105 ml of the binary and quaternary mixtures are sufficient for an extensive VLE characterization.

The RAMSPEQU setup and its integrated workflow enable the characterization of multicomponent VLE while saving significant amounts of substance and laboratory time.

Chapter 1

Introduction

Vapor-liquid equilibrium (VLE) data is at the fundamental core of the chemical industry (Hendriks et al., 2010), as VLE data is crucial for process design, in particular for distillation. In turn, distillation plays a significant role for society: 10 - 15 % of the world energy demand is used for thermal separation processes, foremost for distillation (Sholl and Lively, 2016). Still, even for mixtures of industrial importance, VLE data is often scarce (Kontogeorgis and Folas, 2010). This is due to the fact that the measurement of VLE data for mixtures is still expensive (Hendriks et al., 2010; Kontogeorgis and Folas, 2010; Raal and Mühlbauer, 1998). One reason for VLE measurements being expensive is time: VLE measurements usually require a large amount of work. The amount of work rises strongly with the number of components in a mixture (Raal and Mühlbauer, 1998). As a result, multicomponent data is particularly scarce: the Dortmund Data Bank (DDB), the largest factual data bank for thermophysical properties, contains VLE data for 24425 mixtures. Out of these 24425 mixtures only 155 are quaternary mixtures (DDBST, 2018b). However, VLE data of multicomponent mixtures is the most required property data in the chemical industry (Hendriks et al., 2010). Therefore, industry has a strong interest in faster measurements. Faster measurements can reduce expensive lab time and facilitate characterization of multicomponent systems.

VLE measurements are also expensive because commercial equipment usually needs large amounts of substances. The most commonly employed type of equipment today are dynamic VLE stills. One typical example is the FISCHER® LABODEST® VLE 602. A single filling of the setup requires about 80 ml of substances. Such large amounts of substances are particularly problematic if novel substances are of interest, e.g. tailor-made biofuels. Novel substances are often costly and, even worse, usually only available in small amounts. Consequently, the characterization of the VLE of novel substances and in particular their mixtures is either very costly or can even not be realized at all using conventional techniques.

Biofuels are mentioned in this context as they motivate this work: this work was funded by the Cluster of Excellence Tailor-Made Fuels from Biomass (TMFB). TMFB's focus is on second generation biofuels that are derived from lignocellulosic biomass and therefore restrict competition with food production (Sims et al., 2010). The vision of TMFB is a fuel design process that enables finding the optimum of biofuel production and combustion. For this purpose, VLE are important twofold: first, VLE data is at the basis of process design, and second, to enable a better understanding of the combustion characteristics, VLE data is required for the modeling of fuel sprays and mixture formation in combustion engines (Sirignano, 2014). To find the optimum of biofuel production and combustion, a large number of known and novel substances, and their mixtures has to be studied. For this purpose, screening measurements are desirable. Thus, enabling the rapid characterization of multicomponent VLE based on small amounts of substances is what we[1] aim for in this work.

In addition to novel substances being expensive and possibly hardly available, there is usually no data related to the safety of the substances (e.g., toxicity). Consequently, novel substances have to be considered as potentially harmful. Handling smaller amounts of harmful substances reduces risk.

The extensive laboratory time and large amounts of substances currently necessary for VLE measurements motivate us to design, build and validate an efficient setup for the characterization of VLE.

1.1 Structure of this Thesis

In this work, we introduce a novel workflow and milliliter-scale setup for the efficient characterization of vapor-liquid equilibria. Setup and workflow aim at fast and substance-saving measurements.

Chapter 2 introduces the most commonly employed models for phase equilibrium calculations from literature and illustrates why we choose the perturbed-chain polar statistical associating fluid theory (PCP-SAFT) equation of state (EOS) for phase equilibrium calculations in this work. Classes of experimental VLE characterization are briefly discussed based on a classification from literature. Raman spectroscopic techniques for VLE characterization from literature are presented in more detail. The presentation of VLE characterization equipment from literature reveals some limitations. Section 2.3 discusses how these limitations are tackled by the novel setup and summarizes the contribution of this thesis.

[1]This thesis is written in the *pluralis modestiae* to avoid the excessive use of passive voice.

In Chapter 3 we employ an established setup (Cailletet setup) to characterize phase equilibria of two promising binary biofuel blends at high pressure. The established setup serves as a state of the art reference method.

In Chapter 4, we introduce and validate a Raman spectroscopic milliliter-scale setup and its workflow for the efficient characterization of VLE: RAMSPEQU (**Ram**an **S**pectroscopic **P**hase **Equ**ilibrium Characterization). The RAMSPEQU setup is applied for the characterization of a quaternary model biofuel and its binary subsystems in Chapter 5. Finally, in Chapter 6 the work is summarized and perspectives for future work are given.

Chapter 2

State of the Art: Measurement and Modeling of Vapor-liquid Equilibria

This Chapter introduces the essentials of vapor-liquid equilibrium characterization both from a theoretical and practical point of view. In Section 2.1, thermodynamic models and their fundamentals are discussed. Historically, there are many approaches for phase equilibrium calculations. Here, we introduce the most commonly employed thermodynamic models for inter- and extrapolation of VLE data for process design. We show that PCP-SAFT is well-suited for our purposes. Additionally, the essentials of PCP-SAFT as employed in this work are discussed (see Section 2.1.1).

In Section 2.2, experimental VLE measurements are classified and the basic equipment types are briefly introduced. More detailed insight is provided on VLE measurement equipment and methods using Raman spectroscopy. Based on the limitations revealed in Section 2.2, the idea for a novel VLE measurement setup and workflow are outlined in Section 2.3 which summarizes the contribution of this thesis.

Parts of this Chapter are reproduced by permission of Elsevier from:

Liebergesell, B., Flake, C., Brands, T., Koß, H.-J., and Bardow, A. (2017) A milliliter-scale setup for the efficient characterization of isothermal vapor-liquid equilibria using Raman spectroscopy, *Fluid Phase Equilibria*, 446:36-45.

and

Liebergesell, B., Brands, T., Koß, H.-J., and Bardow, A. (2018) Quaternary isothermal vapor-liquid equilibrium of the model biofuel 2-butanone + n-heptane + tetrahydrofuran + cyclohexane using Raman spectroscopic characterization, *Fluid Phase Equilibria*, 472:107-116.

2.1 Vapor-liquid equilibrium and thermodynamic models

In this Section, popular thermodynamic models and their fundamentals are introduced based on the presentation of Dohrn (1994), Pfennig (2004), and Kontogeorgis and Folas (2010).

For the mathematical description of vapor-liquid phase equilibria, thermodynamic models relate pressure p, temperature T, liquid phase composition x_i and vapor phase composition y_i. Two phases α and β are in equilibrium if the following conditions are met:

$$T^{\alpha} = T^{\beta}, \tag{2.1}$$

$$p^{\alpha} = p^{\beta}, \tag{2.2}$$

$$\mu_i^{\alpha} = \mu_i^{\beta}. \tag{2.3}$$

Equality of chemical potentials μ_i of all species i in all phases (Equation 2.3) is the starting point for equilibrium calculations. The chemical potential μ_i is equal to the partial molar property of a thermodynamic potential of a system (see Equations 2.4 and 2.5). Thermodynamic models are usually based on either the Gibbs free energy G or Helmholtz free energy A:

$$\mu_i = \left(\frac{\partial G(T,p,x_j)}{\partial n_i}\right)_{T,p,n_{j\neq i}} \tag{2.4}$$

$$\mu_i = \left(\frac{\partial A(T,V,x_j)}{\partial n_i}\right)_{T,V,n_{j\neq i}} \tag{2.5}$$

Both thermodynamic potentials - Gibbs free energy G and Helmholtz free energy A - depend on experimentally accessible variables (T,p,x_i or T,V,x_i) only. The two largest families of thermodynamic models are the **G^{E}-models**, that are based on the Gibbs free energy G, and **equations of state (EOS)** that lead to an expression for the Helmholtz free energy A. The main difference between these two families of thermodynamic models is that G^{E}-models deal with liquid phase non-idealities only while EOS usually consider the non-idealities of all fluid phases in equilibrium.

G^{E}-models

G^{E}-models describe the excess Gibbs energy. The excess Gibbs energy is the difference in Gibbs energy of a real mixture compared to the ideal mixture of real components (Pfennig, 2004). G^{E}-models focus on the liquid phase: the excess Gibbs energy G^{E} is modeled as function of temperature and liquid phase composition. Typically, G^{E}-models do not consider pressure dependency. For practical applications, Equation 2.3 can be rewritten resulting in the well-known γ-φ-approach:

$$\gamma_i \cdot x_i \cdot p_i^s = \left(\frac{\varphi_i}{\varphi_i^0}\right) \cdot y_i \cdot p \tag{2.6}$$

where γ_i is the activity coefficient of component i in the liquid phase, φ_i^0 and φ_i are the fugacity coefficients of pure component i and the fugacity coefficient of component i in a mixture respectively, p_i^s is the vapor pressure of pure component i and p the total pressure in the system. The liquid phase and its activity coefficient γ_i on the left side of Equation 2.6 are considered by the G^{E}-model. The fugacity coefficients φ_i and φ_i^0 describing vapor phase non-idealities on the right side of the equation can be calculated from an EOS-model. Alternatively, if ideal gas behavior can be assumed, the influence of the fugacity coefficient cancels ($\varphi_i = \varphi_i^0 = 1$). If fugacities from an EOS are considered, different types of models are combined and special attention has to be paid to combine the models consistently.

The most successful G^{E}-models are based on the local composition approach: the non-random two-liquid model (NRTL) (Renon and Prausnitz, 1968) and the universal quasi-chemical model (UNIQUAC) (Abrams and Prausnitz, 1975). Local composition models account for local interactions between molecules in a mixture. NRTL is probably the most widely used tool in chemical industry today (Hendriks et al., 2010). For binary systems, three parameters are sufficient. Multicomponent systems are readily described. However, NRTL lacks an entropic term. This entropic term is accounted for in UNIQUAC. Therein, size and shape of the molecules are considered, which is particularly significant if asymmetric systems are studied. For both, NRTL and UNIQUAC, pressure and temperature have to be significantly below the critical values. For a satisfying description of the temperature dependence, temperature-dependent parameters are usually necessary. Parameters for a plethora of systems are listed in databases (e.g., relative van der Waals volume r and relative van der Waals surface q for UNIQUAC for 11418 pure components in the DDB (DDBST, 2018a), binary NRTL and UNIQUAC parameters for $>$16000 binary systems in 35 volumes of Chemistry Data Series, e.g. Gmehling et al. (1991)). The availability of parameters is one of the main reasons for the overwhelming popularity of NRTL and UNIQUAC until today.

NRTL and UNIQUAC parameters are commonly adjusted to experimental data. As a result, the experimental data are described by the models (descriptive models). In addition to such descriptive models, there are predictive G^{E}-models. Two fundamentally different predictive approaches are probably the most popular ones today (Hendriks et al., 2010): UNIFAC (UNIQUAC Functional-group Activity Coefficients) and COSMO-RS (Conductor-like Screening Model for Real Solvents). UNIFAC (Fredenslund et al., 1975) predicts parameters from group contributions that are fitted to large amounts of experimental data. In contrast, COSMO-RS (Klamt, 1995) does not use any experimental data for parameter fitting but relies on quantum chemically generated charge density surfaces to describe molecules and their interactions. For most substances, UNIFAC is the more accurate model, particularly if groups and interactions have been parametrized using similar substances (Xue et al., 2012). However, COSMO-RS has a wider range of applicability and exhibits higher predictability as there are no restrictions to certain groups.

Equations of State

In contrast to G^{E}-models, EOS deal with both the vapor and the liquid phase. Consequently, modeling phase equilibria consistently is easier than for G^{E}-models. For EOS calculations, Equation 2.3 reduces to

$$f_i^{\text{liquid}} = f_i^{\text{vapor}} \tag{2.7}$$

with f_i^{phase} the fugacity of component i in the liquid, respectively vapor phase (Lewis, 1901). Fugacity (*fugare*, lat. *to flee*) can be interpreted as the tendency of a component to leave a phase. Comprehensibly, the fugacities have to be equal in both phases in equilibrium (isofugacity criterion). The fugacity f is defined as

$$f_i = \varphi_i x_i p. \tag{2.8}$$

Insertion of Equation 2.8 into Equation 2.7 leads to the well-known $\varphi - \varphi$-approach:

$$x_i^{\text{liquid}} \varphi_i^{\text{liquid}} = x_i^{\text{vapor}} \varphi_i^{\text{vapor}}. \tag{2.9}$$

The first EOS that was able to describe the phase behavior of fluids qualitatively correct (e.g., condensation phenomenon) is the van-der-Waals (vdW) EOS (van der Waals, 1873). The vdW EOS relies on two parameters describing the attractive (a)

and repulsive (b) interactions of molecules (see Equation 2.10).

$$p = \frac{RT}{v-b} - \frac{a}{v^2} \tag{2.10}$$

The parameters a and b can readily be calculated from critical data. Written in its pressure explicit form, the vdW EOS is cubic with respect to volume v. The vdW EOS and similar equations are therefore commonly known as cubic equations of state. For mixtures, parameters are usually calculated using the one-fluid theory: a single pseudo-component is modeled using mixing rules and combination rules. One-fluid parameters are composition dependent.

Even though it does not lead to satisfying quantitative results, the vdW EOS has been the starting point for a plethora of modified cubic equations of state (400+ published cubic EOS as of 2003 (Valderrama, 2003)). The two most successful modifications are still used in industry today: the Soave-Redlich-Kwong (SRK) EOS (Soave, 1972; Redlich and Kwong, 1949) and the Peng-Robinson (PR) EOS (Peng and Robinson, 1976). The SRK-EOS is essentially a vdW EOS with a temperature dependent attraction term and a slight modification of volume dependency. The original RK EOS aimed for an improvement of the calculation of fugacities (Redlich and Kwong, 1949). The SRK EOS enhances the calculation of vapor pressures significantly compared to the vdW EOS (Soave, 1972). However, the SRK EOS is not able to describe liquid phase densities well. A quantitative description of VLE is not satisfying based the SRK EOS alone. However, the SRK EOS is useful for combination with G^{E}-models. The PR EOS has a further modified volume dependency compared to SRK. The result is a better description of liquid phase densities and the near-critical region. The SRK and PR EOS are successful even in industrial applications, as the description of vapor pressure and fugacity is more important for phase equilibrium calculation than density (Dohrn, 1994). Similar to NRTL and UNIQUAC, the availability of parameters is a main reason for the ongoing popularity of the SRK and PR EOS. Parameters are readily calculated from critical data, which is available from large databases. Alternatively, parameters can be fitted to experimental pVT-data to make the equations more reliable in a certain region of conditions.

Cubic EOS try to capture all relevant molecular interactions using two to three parameters. Commonly, a single parameter is intended to describe all types of attractive interactions (e.g., dispersion, polar interactions). As a result, the influence of temperature on different kinds of interactions cannot be considered separately. However, molecular interactions do in fact show different dependencies on density and temperature. Therefore, it is not surprising that the description of phase equilibria based on cubic EOS cannot be arbitrarily accurate.

More complex EOS exist but do not necessarily lead to better results for phase equilibrium calculations. Additionally, in practical applications, they often prove to be less useful due to their complexity (Dohrn, 1994).

However, a noteworthy exception are multiparameter EOS that describe thermodynamic property data very accurately, e.g., by Span and Wagner (1996) and Span et al. (2000). Such approaches represent even the most accurate experimental data within their experimental uncertainty (Span and Wagner, 1996). Therefore, the EOS are very interesting for applications in industry. However, each equation is highly specialized for a certain substance. The equations describe pure component data and are available for a very limited number of substances only.

In summary, cubic EOS are more useful than G^{E}-models once pressure and temperature tend towards critical conditions. However, critical points are still not described accurately by cubic EOS and usually overestimated. The description of the pVT-behavior of pure substances and mixtures by cubic EOS is not satisfying as soon as strongly non-ideal systems are considered (i.e., polar, associating or strongly asymmetric). The availability of the parameters, even for mixtures, is an important reason for the ongoing popularity of the SRK and PR EOS, as well as the NRTL and UNIQUAC G^{E}-models in industrial applications. The choice of an appropriate model for a certain purpose is an expert's task that strongly depends on conditions and systems under study.

In particular within the last 20 years, EOS calculations improved considerably even for strongly non-ideal mixtures. One reason for this improvement is the appearance of SAFT-type equations (SAFT: Statistical associating fluid theory, original SAFT equation by Chapman et al. (1989)). Since the publication of the original SAFT equation, a plethora of variations has been published. In this work, our focus is on one of the most successful variants: PCP-SAFT (PCP: Perturbed-chain polar, (Gross and Sadowski, 2001; Gross, 2005; Gross and Vrabec, 2006), see Section 2.1.1).

In contrast to cubic EOS, PCP-SAFT considers intermolecular forces independently: repulsion, dispersion, association, dipole-dipole and quadrupole-quadrupole interactions are considered to be contributions to the modeled residual Helmholtz energy A^{res} (see Section 2.1.1). Thus, PCP-SAFT exhibits a strong physical background. The enhanced physical background adds mathematical complexity. As a result, PCP-SAFT is more computationally demanding than cubic EOS. However, computing power is not as much of an issue as it used to be some decades ago.

Critical points are still not described quantitatively by SAFT-type EOS (Tang and

Gross, 2010). However, PCP-SAFT leads to considerably better results than cubic EOS for both, the vicinity of the critical point as well as low temperature densities (Gross and Vrabec, 2006). Furthermore, PCP-SAFT can describe the phase behavior of polar molecules and their mixtures quantitatively. Due to the physical background, the estimation of parameters based on quantum mechanics is possible (van Nhu et al., 2008). Temperature-dependent parameters are usually not necessary. Instead, the phase behavior can often be extrapolated with respect to temperature and pressure (Gross and Vrabec, 2006).

PCP-SAFT's ability to describe polar molecules and their mixtures quantitatively is of importance for this work. Herein, second generation biofuels and their blends are of interest. Second generation biofuels are derived from lignocellulosic biomass. Thus, a high oxygen content is expected, leading to a high polarity of the substances. Furthermore, we consider both high pressure (Chapter 3) and low pressure conditions (Chapters 4 and 5) in this work. PCP-SAFT has favorable abilities for polar substances at both high and low pressure conditions. Due to its versatility, we choose PCP-SAFT for thermodynamic modeling in this work.

2.1.1 The PCP-SAFT equation of state

PCP-SAFT models the residual Helmholtz energy (A^{res}) as a sum of physical contributions (Gross and Sadowski, 2001, 2000; Gross and Vrabec, 2006; Gross, 2005):

$$A^{\text{res}} = A^{\text{hs}} + A^{\text{chain}} + A^{\text{disp}} + A^{\text{DM}} + A^{\text{QM}}. \tag{2.11}$$

Pure components are modeled as a chain of tangentially connected, spherical segments with an effective diameter σ_i. The number of segments is described by the parameter m_i. The effective diameter σ_i and the number of segments m_i are accounted for in the hard sphere contribution (A^{hs}) and the contribution for the chain formation from individual segments (A^{chain}) (Gross and Sadowski, 2001). Additionally, PCP-SAFT accounts for the following contributions: dispersion interactions (A^{disp}, parameter ϵ_i) (Gross and Sadowski, 2001), dipole-dipole interactions (A^{DM}, parameter μ_i) (Gross and Vrabec, 2006), and quadrupole-quadrupole interactions (A^{QM}, parameter θ_i) (Gross, 2005). The model used in this work does not consider association, since only non-associating compounds are studied. For a more detailed discussion of the PCP-SAFT model, we refer the reader to the original publications (Gross and Sadowski, 2001, 2000; Gross and Vrabec, 2006; Gross, 2005).

All equation of state (EOS) calculations are performed using in-house software. The objective function for parameter estimation is a weighted least squares minimization of all dependent values (i.e., pressure p, density ρ), weighted by their experimental measurement uncertainty. For each component studied, we fit a set of pure component parameters m_i, σ_i and ϵ_i to one liquid density and all vapor pressure data obtained for the respective pure component. This experimental data is choosen for fitting of the pure component PCP-SAFT parameters based on the work of Albers and Sadowski (2011). Albers and Sadowski recommend using at least one liquid density and at least two vapor pressure values as the minimal data set for safe parameter estimation. Furthermore, Albers and Sadowski report that the influence of vapor pressure data is larger, not only on parameter estimation, but also on the subsequent modeling results than the influence of liquid density data. Therefore, we use all of the measured vapor pressures to estimate the pure component parameters. The parameters for the dipole (μ_i) and quadrupole (θ_i) contributions are not fitted but determined based on quantum mechanical calculations as described in detail by van Nhu et al. (2008).

Mixture properties are calculated using conventional one-fluid mixing rules (Gross and Sadowski, 2001, 2000). We calculate the one-fluid mixture dispersion parameter ϵ_{ij} as follows:

$$\epsilon_{ij} = \sqrt{\epsilon_i \epsilon_j} \cdot (1 - k_{ij}) \,. \tag{2.12}$$

In this work, binary interaction parameters k_{ij} are either set to zero or fitted to mixture data.

2.2 Measuring vapor-liquid equilibria

First well-documented methods for the characterization of VLE are known from the end of the 19th century (Carveth, 1899). Since then, numerous variants of a few basic equipment types were developed. Extensive reviews on experimental equipment and procedures can be found in the literature for low pressure (Raal and Ramjugernath, 2005; Raal and Mühlbauer, 1998; Abbott, 1986) and high pressure measurements (Fonseca et al., 2011; Dohrn et al., 2012). Herein, the most commonly used types of equipment will be introduced based on the presentation of Raal and Mühlbauer (1998) and Raal and Ramjugernath (2005). A more detailed discussion is presented for VLE equipment using Raman spectroscopy. Limitations of current VLE setups are revealed that motivate the idea for the novel VLE equipment and measurement procedure that is the main topic of this thesis.

Vapor-liquid equilibrium measurements can be classified as shown in Figure 2.1 according to Dohrn et al. (2012).

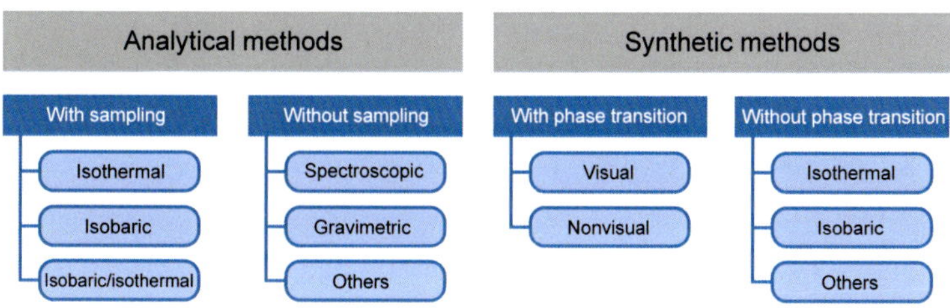

Figure 2.1: Classification of experimental methods for VLE according to Dohrn et al. (2012).

The classification by Dohrn et al. (2012) follows characteristic error sources of the respective types of equipment. Therefore, the classification does not necessarily cover all details of a measurement method. For instance, in contrast to analytical methods based on sampling, there is no discrimination between spectroscopic techniques (e.g., isothermal or isobaric). Still, we deem the classification by Dohrn et al. (2012) to be the most fitting approach for our purposes.

According to Figure 2.1, there are two main categories of VLE characterization: analytical and synthetic methods. Synthetic methods rely on samples with known composition. The composition is known as samples are precisely prepared ("synthesized"). As a result, sample preparation is usually complicated. Degassing is crucial. Substances either have to be degassed prior to preparation of a mixture, resulting in complicated preparation of the mixture, or lengthy freeze/thaw cycles (see Section 3.2) have to be used for degassing of the mixture to avoid any changes in composition. Degassed substances have to be transferred to an equilibrium cell without any contact with the atmosphere. Nevertheless, synthetic methods are the most commonly employed method for VLE measurements, particularly at high pressures (Fonseca et al., 2011), as complex analytics and sampling issues are avoided. Synthetic methods can further be divided into methods with phase transition or without phase transition. In methods without phase transition, phase compositions are calculated based on known phase densities and the volume of the equilibrium cell. Synthetic methods without phase transition are not frequently used. Instead, the largest share of measurements is conducted using synthetic methods with phase transition and visual detection. We use such a synthetic method with phase transition and visual detection in Chapter 3. Pressure or temperature is changed till a new phase appears or one of the phases disappears. Usually, pressure is changed until the vapor phase just disappears. Thus, the liquid phase is observed directly at the point of phase transition (here: bubble point). As a result, the liquid phase composition is equal to the overall composition. Hence, a synthetic method with phase transition commonly gives access to one pTx-data point per experiment. However, it is possible to use a single filling of a setup to generate multiple data points, e.g., if temperature is changed and pressure adjusted accordingly (as done in Chapter 3). In general, synthetic methods do not provide the composition of the phases from independent measurements but either get the composition directly from synthesis (with phase transition) or infer compositions from computation (without phase transition).

In contrast, analytical methods provide independent information on the phase composition by measuring the composition of at least one phase. Commonly, phase compositions are determined by sampling. According to Figure 2.1, analytical methods with sampling are divided into "Isothermal", "Isobaric" and "Isobaric/Isothermal" methods. Most commonly, isothermal measurements are conducted using static methods (see below) and isobaric measurements are conducted using dynamic methods. From here on, we use the terms "static" and "dynamic", as they avoid ambiguity in the description of measurement setups. In **dynamic methods**, substances are boiled using a heating element and vapor and/or liquid phase pass through an equilibrium cell. After passing through the equilibrium cell, the condensate of both phases is separately led back to a mixing chamber. While phases are flowing back to the mixing chamber, samples can easily be taken using separate sampling points for each phase (Raal and Ramjugernath, 2005). This easy access to samples is one reason for dynamic methods being widely applied, particularly for low pressure VLE measurements. A typical example is the FISCHER® LABODEST® VLE 602 equilibrium still. However, due to their design, dynamic methods lead to gradients in temperature and concentration in the system which are avoided by **static methods**. Static methods are therefore considered to yield "the truest equilibrium" according to, e.g., Raal and Ramjugernath (2005). Static methods commonly give access to isothermal VLE data. An equilibrium cell contains a liquid mixture that is agitated to achieve equilibrium with its vapor at a fixed temperature (Raal and Mühlbauer, 1998). However, for practical implementation, static techniques are rather cumbersome according to Raal and Mühlbauer (1998). In particular, pure degassed liquids have to be added into an evacuated equilibrium cell (Raal and Ramjugernath, 2005). Degassing is crucial but unfortunately very time consuming (usually 2 to 24 hours) (Raal and Ramjugernath, 2005; Battino et al., 1971; van Ness and Abbott, 1978) and requires significant amounts of substance (Narasigadu, 2011).

A static method can also employ an analytical method to determine a complete $pTxy$-data set. However, if the analytical method requires sampling, it will give rise to errors, e.g., due to a change of composition or pressure, consequently shifting the equilibrium. The sampling error can be kept small if the relative sampling volume is small. A small relative sampling volume can be achieved by large cell volumes (Peper and Dohrn, 2012). However, large cell volumes can be a major cost factor due to the amount of substances necessary for phase equilibrium characterization. A large amount of substances is particularly critical if novel substances are studied. In addition, large cell volumes cause long equilibration times as the system responds slowly (Peper and Dohrn, 2012).

Even if large cell volumes are used to minimize the sampling error on the equilibrium, sampling errors can still occur due to changes of the sample itself: parts of the sample might be retained in the sampling line or components might be changed chemically during sampling or due to reaction with air. Loss of components from the sample also has to be considered as a possible error source (Peper and Dohrn, 2012).

In order to avoid sampling and all its associated issues, the compositions of the phases can be determined non-invasively. The non-invasive measurement of composition can be conducted using gravimetric methods. However, gravimetric methods are typically restricted to a very limited group of systems containing non-volatile substances such as ionic liquids or polymers (Fonseca et al., 2011). Gravimetric methods are commonly applied for binary mixtures, as considerable modifications are necessary for the analysis of ternary mixtures already (Tanbonliong and Prausnitz, 1997). Alternatively, compositions can be determined non-invasively by spectroscopy as the analytical tool. As it avoids sampling, spectroscopy enables composition measurements using small cell volumes. Spectroscopic techniques have already been applied for non-invasive analysis of VLE, foremost infrared (Alsmeyer et al., 2002; Borges et al., 2015), and Raman spectroscopy (Kaiser et al., 1992; Stratmann and Schweiger, 2002; Adami et al., 2013; Luther et al., 2015). In this work, we focus on Raman spectroscopy for analysis. Raman spectroscopy provides qualitative and quantitative information on all species in a mixture (Pelletier, 2003). Raman-based VLE methods, as the RAMSPEQU setup developed in this work, can be classified as an analytical method without sampling using spectroscopic analysis of composition (see Figure 2.1).

Even though spectroscopic techniques have been used as analytical method for VLE measurements before, the most recent data shows that only 1.1 % of systems were studied using spectroscopic methods (most recent data only available from 2005 - 2008, at high pressure) (Fonseca et al., 2011). The small number of systems studied is at least in part due to the limited number of available spectroscopic VLE setups.

Laboratories that can realize both spectroscopic and phase equilibrium measurements are rare. Still, spectroscopic techniques exhibit advantages over techniques based on sampling. Prior to introducing Raman spectroscopic VLE measurement techniques from literature, we give a brief introduction to the composition measurement using Raman spectroscopy.

Similar to chromatographic techniques, which are commonly used in combination with sampling, calibration is necessary for the measurement of phase compositions with Raman spectroscopy. In general, the amount of substance n_i of a component i is proportional to its Raman signal S_i (see Equation 2.13):

$$n_i = k_i \cdot S_i. \tag{2.13}$$

The calibration constant k_i depends on (1) the properties of the component (i.e. Raman scattering cross section) and (2) the measurement setup and conditions.

From Equation 2.13 we get for the composition in terms of the mole fraction x_i of a binary mixture of components i and j:

$$x_i = \frac{n_i}{n_i + n_i} = \frac{k_i \cdot S_i}{k_i \cdot S_i + k_j \cdot S_j} \tag{2.14}$$

$$\text{or} = \frac{k_{ij} \cdot S_i}{k_{ij} \cdot S_i + S_j} \text{ with } k_{ij} = \frac{k_i}{k_j}. \tag{2.15}$$

The calibration constants k_i and k_j can be determined independently (see Equation 2.14), based on measurements of pure components. The independent determination of calibration constants is known as *absolute calibration*. Alternatively, a single calibration constant k_{ij} can be determined for the mixture of i and j based on mixtures of known composition (see Equation 2.15). The calibration based on mixtures of known composition is known as *relative calibration*. For an absolute calibration, there is no need to prepare mixtures of known composition. However, measurement setup and conditions can change from measurement to measurement (e.g., fluctuations in laser light irradiation) and affect the calibration constants from absolute calibration separately. In contrast, for relative calibration, mixtures of known composition have to be prepared. However, using a relative calibration cancels effects caused by measurement setup and conditions. As a result, calibration constants from relative calibration are more reliable as changes in measurement setup and conditions do not affect the measurement of composition.

First Raman-based VLE experiments were published by Kaiser et al. (1992). Kaiser et al. study the phase equilibrium of cyclohexane + toluene at T = 323 - 343 K at a pressure of p = 12 - 73 kPa. Similar measurement conditions are studied in Chapters 4 and 5 using our novel RAMSPEQU setup. Kaiser et al. (1992) use Raman spectroscopy to measure the phase composition in both phases. They use an absolute calibration procedure in the liquid phase. For concentration measurements, they assume that the intensity of the mixture spectra is the weighted sum of the intensities of the pure components. Kaiser et al. (1992) perform calibration measurements for the liquid phase only. A calibration factor for the vapor phase is determined using the liquid phase calibration factor and multiplying it with a correction factor from optical theory. Evidently, as the phase equilibrium data is in agreement with data from literature, this calibration procedure for the liquid and vapor phase is sufficient for the system in question. However, a different calibration procedure is chosen by the same group for another system 10 years later (see below, study by Stratmann and Schweiger (2002)).

Kaiser et al. (1992) state a disadvantage of their technique: they used a time-consuming photon-counting technique (photomultiplier) for the measurement of Raman spectra. The measurement of a single vapor phase Raman spectrum took 85 minutes (Kaiser et al., 1992). The volumetric size of the equilibrium cell employed in their work is not mentioned. However, Kaiser et al. conclude that the most important advantage of their technique is the non-invasive measurement of composition that "*could* be applied to very small samples" (Kaiser et al., 1992).

Further Raman-based studies focus on high pressure VLE (Stratmann and Schweiger, 2002; Adami et al., 2013; Luther et al., 2015). Stratmann and Schweiger (2002) study the phase equilibrium of ethanol + carbon dioxide at T = 313.15 K and pressures of p = 0.5 - 7.9 MPa. The equilibrium cell in their study has a volume of 272 ml. For composition measurement, the liquid phase is calibrated based on the Raman spectrum of a mixture of known composition (relative calibration). In contrast to Kaiser et al. (1992), Stratmann and Schweiger (2002) calibrate the vapor phase based on measurements instead of relying on a correction factor from theory. In the vapor phase, an absolute calibration procedure is used: the equilibrium cell is loaded with gaseous ethanol respectively carbon dioxide at pressures below their saturation pressures, one after the other. Consequently, condensation is avoided. However, as an absolute calibration is used, any effects that might cause fluctuations of the absolute Raman signal have to be strictly avoided. Stratmann and Schweiger (2002) report finding a systematic deviation in the vapor phase composition compared to literature data. They state that the deviation can be eliminated by numerically altering the calibration factor. As the vapor phase data would agree well with literature if another calibration factor was found, the deviation could be explained with a single erroneous calibration measurement, e.g., caused by a change of the measurement setup or conditions. Thus, the observed deviation might have been avoided by employing a relative calibration procedure in the vapor phase as well.

Adami et al. (2013) study the phase equilibrium of acetone + water + carbon dioxide at T = 313 - 353 K at a pressure of p = 10 MPa. They use a variable volume equilibrium cell with a volume of 25 - 50 ml. Adami et al. (2013) use the same calibration factors for the Raman spectroscopic measurement of composition in both phases. They employ a relative calibration using several mixtures of known composition. Adami et al. (2013) do not state in which phase the calibration measurements are performed. Acetone, water and carbon dioxide exhibit strong molecular interactions (e.g., hydrogen bonding). In general, molecular interactions can cause nonlinear changes in Raman spectra and thereby complicate their interpretation. The nonlinear changes in the Raman spectra can be significant and consequently either lead to erroneous results or have to be accounted for explicitly. Adami et al. (2013) state that in their study, spectral changes due to interactions are only influenced by composition. For spectral evaluation, they use an especially designed sequence of background correction, normalization, and substraction of certain signals of mixtures of known composition. Consequently, calibration and signal evaluation procedure are restricted to the respective substances and conditions.

Luther et al. (2015) also study the phase equilibrium of acetone + water + carbon dioxide (or nitrogen) at temperatures of T = 303 - 333 K and pressures of p = 6 - 10 MPa. They make progress in the rapid determination of high pressure VLE data by combining Raman spectroscopy with microfluidics. A relative calibration procedure is used. For spectral evaluation, a deconvolution technique (Schuster et al., 2014) is employed that considers spectral changes due to molecular interactions. Luther et al. (2015) state that, in comparison to literature data, a systematic deviation can be observed in all measured vapor phase compositions due to a liquid film that surrounds the vapor phase and emits an interfering Raman signal. Furthermore, the authors state that their method would require considerable modifications of the Raman excitation and detection system for the characterization of vapor phase compositions at low pressure. At high pressure, the density in the vapor phase is higher, resulting in a facilitated detection of the vapor phase Raman signal. At low total pressures, the density in the vapor phase is very small. Therefore, characterization of the vapor phase is more challenging.

2.3 Contribution of this Thesis

In Section 2.2, previous VLE characterization methods are discussed. Despite numerous efforts regarding the design of VLE measurement equipment, a gap exists: a substance-saving and fast method for the efficient characterization of VLE is not available. This motivates the idea for the central scientific contribution of this thesis: the novel measurement setup and procedure RAMSPEQU.

Binary high pressure VLE using a Cailletet setup

First, the well-established Cailletet setup is employed to characterize the VLE of two promising binary biofuel blends at high pressure and temperature (see Chapter 3). The introduced Cailletet setup serves as a state of the art reference setup.

RAMSPEQU: A milliliter-scale setup for VLE

The Cailletet method employed in Chapter 3 is very well-engineered and exhibits remarkable accuracy. However, this comes at the cost of significant complexity. To avoid excessive infrastructure and to simplify the handling of setup and samples, we introduce the milliliter-scale setup RAMSPEQU (**Ram**an **S**pectroscopic **P**hase **Equ**ilibrium Characterization) for the efficient characterization of VLE (see Chapter 4). The designed setup and measurement procedure aims at easy handling, and a substance-saving and fast characterization of complete VLE-data sets ($pTxy$).

We employ an analytical method without sampling: independent composition information of both phases from Raman spectroscopy avoids complicated sampling procedures. The non-invasive technique enables studying small samples in short time. A novel integrated workflow combines calibration and phase equilibrium characterization to save further time and substance. The setup is intended for VLE measurements at low pressure. Low pressure conditions are less demanding regarding the measurement setup itself, but more challenging for the spectroscopic characterization of vapor phase compositions.

The RAMSPEQU setup and workflow are presented in Chapter 4. Pure component vapor pressures and binary VLE data are measured and compared to literature data for validation.

VLE of a quaternary model biofuel using RAMSPEQU

Mixtures containing only two components rarely depict industrial applications. However, a "reasonably complete" characterization of VLE with more than two components requires extensive experimental effort (Raal and Mühlbauer, 1998). To reduce the experimental effort, an efficient measurement method seems particularly suited for multicomponent VLE characterization. We apply the RAMSPEQU setup for the efficient VLE characterization of a quaternary model biofuel and its six binary subsystems (see Chapter 5). PCP-SAFT calculations are used to compare our binary data to experimental data from literature, and to assess the quality of the quaternary VLE data.

Chapter 3

Binary high pressure VLE using a Cailletet setup

3.1 Introduction

This Chapter presents new VLE data of two promising biofuel blends at high pressure. The VLE data are collected using a well-established measurement method: a Cailletet setup. The Chapter contributes to the thesis, as a state of the art measurement method is introduced to serve as a reference method.

The first blend that is studied is a model fuel for spark-ignition engines composed of 2-methylfuran (2-MF) and *iso*-octane. 2-MF is readily produced from biomass (Lange et al., 2012). 2-MF reduces hydrocarbon emissions and increases efficiency in comparison to model fuels for spark-ignition engines (Thewes et al., 2011). The second blend is a model fuel for compression-ignition engines composed of 2-methyltetrahydrofuran (2-MTHF) + di-*n*-butyl ether (DNBE). Both 2-MTHF and DNBE can be derived from biomass (Geilen et al., 2010; Kim et al., 2013; Ezeji et al., 2004; Bond-Watts et al., 2011; Serrano-Ruiz et al., 2010). 2-MTHF + DNBE is a promising biofuel blend as little to no soot formation can be observed during its combustion in a compression-ignition engine (Janssen et al., 2011).

Bubble-point pressures for both binary mixtures are collected over a wide range of temperature and pressure (T = 353 - 508 K, p = 0.16 - 3.7 MPa). Additionally, vapor pressures of all pure components are measured and compared to literature data. The experimental data are modeled using the PCP-SAFT EOS. We compare predictions of binary phase behavior from pure component parameters to experimental data.

The experimental setup and procedure is presented in Section 3.2. Experimental and modeling results are discussed in Section 3.3. Finally, the results are summarized and conclusions are drawn in Section 3.4.

Major parts of this Chapter are reproduced by permission of Elsevier from:

Liebergesell, B., Kaminski, S., Pauls, C., de Loos, Th. W., Vlugt, T. J. H., Leonhard, K., and Bardow, A. (2015) High-pressure vapor-liquid equilibria of the second generation biofuel blends (2-methylfuran + *iso*-octane) and (2-methyltetra-hydrofuran + di-*n*-butyl ether): Experiments and PCP-SAFT modeling, *Fluid Phase Equilibria*, 400:95-102.

Contribution report: Writing the draft, principal author, planning and conducting the experiments, data evaluation.

PCP-SAFT parameter fitting for this Chapter has been performed by S. Kaminski. Parts of the experiments were performed with assistance by E. Straver, M. Ramdin and M. de Groen.

3.2 Experimental setup and procedure

In this Section, the experimental setup and procedures for the measurements of VLE and density conducted in this Chapter are described. All chemicals used in this chapter and their purities are listed in Table A.1.

Bubble-point pressure measurements

Measurements are carried out using the so-called Cailletet apparatus according to the synthetic method. A detailed description of the experimental set-up is given by de Loos et al. (1986). The equipment provides access to temperatures of up to 508 K and pressures of up to 15 MPa. The temperature is measured to an accuracy within 0.01 K using a platinum resistance thermometer (Pt100) and maintained constant to better than 0.03 K during measurements. The pressure is measured using a hydraulic dead weight gauge (Budenberg 480HX, $u(p) = 0.0005\,\text{MPa}$). The error in the mole fraction of the mixtures is estimated to be less than $u(x) = 0.001$. The estimation is based on the accuracy of the balance and empirical values related to the used degassing technique (see below).

The following standard procedure is employed: The top of a Pyrex glass capillary tube (Cailletet tube) is filled with pure component or binary mixtures of known composition. The substances are thoroughly degassed using a freeze/thaw technique. For the freeze/thaw technique, first, the Cailletet tube is connected to a vacuum rack. The vacuum rack is evacuated using a turbomolecular vacuum pump. The sample is frozen inside of the Cailletet tube using liquid nitrogen. Next, vacuum is applied to remove gas from the tube. Subsequently, the sample is thawed and heated up to room temperature again. Residual gas is removed from the liquid phase by stirring using a magnetic stirring ball inside the tube. The sample is once again frozen using liquid nitrogen and the procedure described above is repeated until no escaping of gas is visible during the procedure and the sample can be assumed to be completely degassed. Typically, to completely degas a sample, 5 to 7 freeze/thaw cycles are necessary. Subsequently, the rest of the Cailletet tube is filled with mercury. The mercury serves the purpose of sealing the sample towards atmosphere respectively the hydraulic oil from the dead weight gauge. The Cailletet tube, filled with degassed sample and mercury, is placed into a thermostat bath filled with silicon oil. The temperature is set to a desired value using a temperature controller. The pressure is adjusted via the dead weight gauge until the coexistence of a gaseous and a liquid phase can be observed visually in the tube. Subsequently, pressure is increased carefully, until the disappearance of the last gas bubble is observed visually, indicating the bubble-point pressure.

For *iso*-octane-rich mixtures of *iso*-octane and 2-MF, precipitation of a flaky solid phase is observed during mixing of the pure components. Each mixture, still containing the precipitate, is homogenized by shaking before a sample is taken. Applying the standard sample preparation procedure, the precipitate was no longer observed after the first degassing step using the freeze/thaw technique (see above).

Density measurements

The density is measured with a DMA 4500 M apparatus (Anton Paar GmbH) according to the oscillating U-tube principle. The repeatability of the temperature is specified as 0.01 K, the repeatability of the density measurement as $0.000\,01\,\mathrm{g\,cm^{-3}}$. The setup is validated at (298.15, 313.15, 333.15 and 353.15) K by measuring the certified standards APN 100 und APN 415. Before measurement, the samples are preheated and stirred in vials in a water bath for degassing at 355 K. For the actual measurement, the samples are equilibrated in the DMA 4500 M for several minutes.

3.3 Results and discussion

In this Section, experimental densities (see Subsection 3.3.1) and vapor pressure data (see Subsection 3.3.2) of 2-MF, 2-MTHF, DNBE and *iso*-octane are presented and compared to data from literature. Additionally, we check the thermal stability and the purity of the components. In Subsection 3.3.3, experimental bubble-point pressure data of the two binary biofuel blends 2-MF + *iso*-octane and 2-MTHF + DNBE are presented and compared to modeling results.

3.3.1 Density measurements

The experimental liquid density data for all pure components are listed in Table 3.1. Densities are measured at $T = 295\,\mathrm{K}$. Density measurements are solely performed to estimate PCP-SAFT parameters. The pure component densities measured in this work are in good agreement with the values for 2-MF (Chadha and Tripathi, 1995), 2-MTHF (Vallés et al., 2006), DNBE (Jiménez et al., 1998) and *iso*-octane (Pádua et al., 1996) reported in literature (see Table 3.1).

Table 3.1: Experimentally determined pure component densities (ρ) at 0.1 MPa for different temperatures (T) with the standard uncertainties $u(i)$ and density values from the literature.

substance	T/K	ρ/g cm^{-3}	
		Experimental	Literature
2-MF	295.00	0.9167	
	293.15		0.9148 (Chadha and Tripathi, 1995)
2-MTHF	295.00	0.8516	
	298.15		0.8490 (Vallés et al., 2006)
DNBE	298.00	0.7638	
	298.15		0.7641 (Jiménez et al., 1998)
iso-octane	295.00	0.6904	
	298.15		0.6878 (Pádua et al., 1996)

$u(T) = 0.01\,\mathrm{K}, u(\rho) = 0.000\,01\,\mathrm{g\,cm^{-3}}$

3.3.2 Pure component vapor pressures

The experimental vapor pressure data for 2-MF, *iso*-octane, 2-MTHF and DNBE is listed in Tables A.2 and A.3. The experimental data is compared to literature data in Figure 3.1.

In case of DNBE, bubble-point pressures are only available for lower (414 K) and higher temperatures (511 - 566 K). To still compare our data to the literature data, we extrapolate our measurements using the PCP-SAFT EOS. For this purpose, the PCP-SAFT parameters are estimated based on our own data (see Section 2.1.1 and

Table 3.2). Bubble-point pressures are calculated from PCP-SAFT for all the temperatures reported by Kobe et al. (1956). As measure of quality, we choose the absolute average deviation Δp of experimental and calculated data:

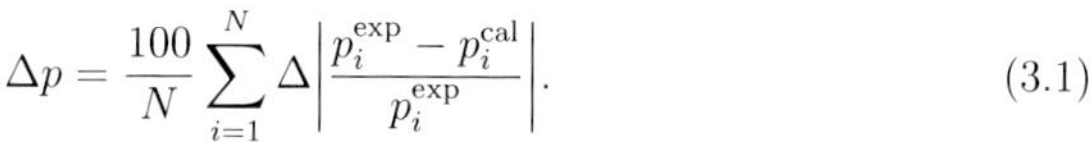

$$\Delta p = \frac{100}{N} \sum_{i=1}^{N} \Delta \left| \frac{p_i^{\text{exp}} - p_i^{\text{cal}}}{p_i^{\text{exp}}} \right| . \tag{3.1}$$

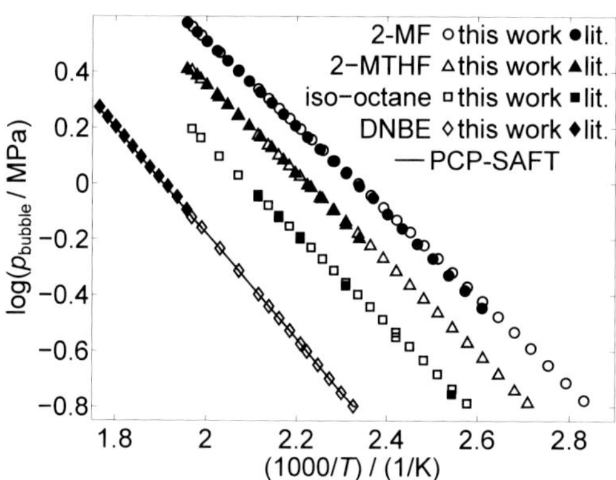

Figure 3.1: Comparison of the experimentally observed vapor pressure of every pure component used in this work (open symbols) to vapor pressure data from literature (2-MF (Kobe et al., 1956), 2-MTHF (Kobe et al., 1956), *iso*-octane (Kay and Warzel, 1951) and DNBE (Kobe et al., 1956), closed symbols). For easier comparability, in case of DNBE, our own data are extrapolated to the conditions from literature using PCP-SAFT (solid line, see Section 2.1.1). The used PCP-SAFT parameters can be found in Table 3.2.

We compare the calculated pressures (p^{cal}) to the experimental literature data (p^{exp}) for all pure components. The pure component vapor pressures collected in this work are in accordance ($\Delta p = 2.4\,\%$) with the data reported in the literature (Kobe et al., 1956; Kay and Warzel, 1951) (see Figure 3.1).

The purity of all components is checked as follows: Vapor pressures are measured both at approximately equal gas and liquid volumes coexisting in equilibrium, and at the bubble-point line, i.e., with zero gas volume. As the vapor pressure of mixtures depends on composition and consequently on the gas-to-liquid phase ratio, a difference in pressure would be expected if impurities were present. Comparison of the pressures at equal phase ratio and at the bubble-point showed no differences larger than the measurement uncertainty. As we found no dependency of the pressure on the phase ratio, we assume the purity of all components to be satisfactory for our needs.

The thermal stability of all pure compounds and several mixtures is checked as follows: Bubble-point pressures are collected using the standard procedure (see Section 3.2) starting from about 0.16 MPa. In this first run of experiments, temperature is increased in 5 K temperature steps up to the maximum temperature of 508 K (see Tables A.2 and A.3). Subsequently, the samples are cooled down to room temperature. 3rd order polynomials are fitted to the collected bubble-point pressures as function of temperature. Then, in a second run of experiments, bubble-point pressures are again collected using the standard procedure but now at temperatures lying in-between the temperatures studied in the first run. Each bubble-point pressure of the second run is then compared to an interpolated polynomial value of the first run. The deviations in terms of pressure are in all cases smaller than the measurement error. Consequently, sufficient thermal stability of all compounds is assumed.

3.3.3 Binary systems

The experimental bubble-point data of the systems 2-MF + *iso*-octane and 2-MTHF + DNBE are listed in Tables A.2 and A.3, respectively. Isotherms of the experimental data are shown in Figures 3.2 and 3.3 together with the respective calculated data from PCP-SAFT.

To obtain a detailed description of the VLE behavior of the binary biofuel blends, bubble-point pressures are collected at equidistant mole fractions of 0.1. For each composition, measurements are started at the temperature corresponding to the lowest pressure accessible with the apparatus (about 0.16 MPa). Starting at this temperature, data is collected in steps of 5 K up to 473 K, and 10 K from 473 K to 503 K, plus a final measurement at the temperature limit of 508 K.

The two binary systems deviate to different extents from ideal mixture behavior: The mixture of 2-MF and *iso*-octane shows a slight positive deviation from ideal behavior. The deviation increases in its extent with decreasing temperature (Figure 3.2). The mixture of DNBE and 2-MTHF deviates in a negative way from ideal behavior (Figure 3.3). The negative deviation is more pronounced with rising temperature.

We calculate the deviations from ideal mixture behavior in terms of pressure for all temperatures where the pure component vapor pressures are available from our measurements. The resulting deviations are $\Delta p_{\text{2-MF}/\textit{iso}\text{-octane,ideal}} = 3.6\,\%$ and $\Delta p_{\text{DNBE/2-MTHF,ideal}} = 4.8\,\%$.

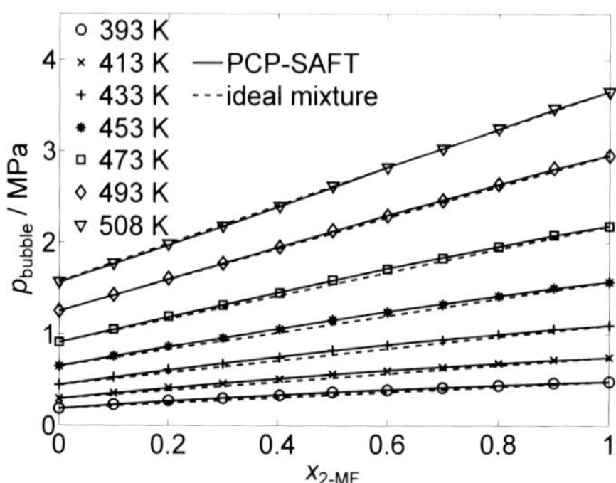

Figure 3.2: Isothermal $p - x$-section (p: bubble-point pressure, x: mole fraction) for the mixture [x 2-MF + (1-x) *iso*-octane]. Symbols are experimental data. The legend links the temperatures to the symbols. The symbols at x = 0 also link the temperature to the corresponding data calculated from PCP-SAFT (solid lines, $k_{2\text{-MF}/iso\text{-octane}} = 8.1 \cdot 10^{-3}$, see Table 3.2 for pure component parameters) and ideal mixture behavior (dotted lines). Data is displayed only at selected temperatures for reasons of clarity.

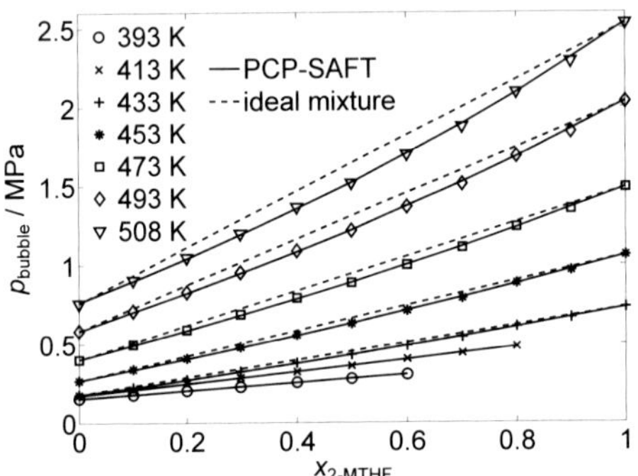

Figure 3.3: Isothermal $p - x$-section (p: bubble-point pressure, x: mole fraction) for the mixture [x 2-MTHF + (1-x) DNBE]. Symbols are experimental data. The legend links the temperatures to the symbols. The symbols at $x = 0$ also link the temperature to the corresponding data calculated from PCP-SAFT (solid lines, $k_{\text{DNBE/2-MTHF}} = -8.0 \cdot 10^{-4}$, see Table 3.2 for pure component parameters) and ideal mixture behavior (dotted lines). Data is displayed only at selected temperatures for reasons of clarity.

The bubble-point pressures of binary mixtures are modeled using the PCP-SAFT EOS (see Section 2.1.1). The pure component parameters used for the calculations are listed in Table 3.2. The pure component parameters are adjusted to one liquid density and all experimental pure component vapor pressures (Figure 3.1, Table 3.1, A.2 and A.3).

Table 3.2: Pure component PCP-SAFT parameters for the calculation of bubble-point pressures. Parameters (m_i, σ_i, ϵ_i) are adjusted to one liquid density (Table 3.1) and the vapor pressure values shown in Figure 3.1 and Tables A.2 and A.3. Dipole and quadrupole parameters (μ_i, θ_i) are calculated as described in detail by van Nhu et al. (2008).

Component	m	σ/Å	(ϵ/k_B) / K	μ/D	θ/DÅ
2-MF	2.4242	3.6613	263.503	0.634	6.785
2-MTHF	2.7191	3.6567	265.094	1.575	4.405
iso-octane	2.9124	4.1988	258.601	0.121	0.707
DNBE	3.7452	3.9221	252.893	3.276	1.098

Binary interaction parameters k_{ij} for the dispersion interactions are fitted to all mixture bubble-point pressures for each system. The resulting binary interaction parameters are $k_{\text{2-MF}/iso\text{-octane}} = 8.1 \cdot 10^{-3}$ and $k_{\text{DNBE/2-MTHF}} = -8.0 \cdot 10^{-4}$. These very small binary interaction coefficients show that the geometric mean rule (setting $k_{ij} = 0$) already yields a suitable estimate for the mixture dispersion parameter ϵ_{ij} in both cases.

As shown in Figures 3.2 and 3.3, the model shows very good agreement with the experimental data. The absolute average deviations of experimental and modeled bubble-point pressures are $\Delta p_{\text{2-MF}/iso\text{-octane}} = 0.68\,\%$ and $\Delta p_{\text{DNBE/2-MTHF}} = 0.61\,\%$. The good agreement and the small Δp-values indicate a high quality of the experimental data under the assumption that there are no systematical errors in the measurements.

For comparison, calculations are performed with a k_{ij} value set to zero for both systems. In this case, PCP-SAFT predicts the mixture behavior from pure component data only. The resulting deviations are $\Delta p_{\text{2-MF}/iso\text{-octane}} = 2.60\,\%$ and $\Delta p_{\text{DNBE/2-MTHF}} = 0.65\,\%$. Surprisingly, the unusual negative deviation from ideal behavior in case of the DNBE/2-MTHF mixture is predicted even better than the slight positive deviation from ideal behavior of 2-MF/*iso*-octane.

3.4 Conclusion

Bubble-point pressures of novel biofuel blends are collected to close an existing gap for vapor-liquid equilibrium data. We conduct measurements at $T = 353$ - $508\,\mathrm{K}$ for 2-MF/*iso*-octane and at $T = 373$ - $508\,\mathrm{K}$ for DNBE/2-MTHF. The bubble-point pressures are obtained using the synthetic Cailletet method. Experimental data for pure components is in good agreement with reference data. The system composed of 2-MF/*iso*-octane shows a slight positive deviation from ideal mixture behavior. For the DNBE/2-MTHF mixture, a more pronounced negative deviation from ideal mixture behavior is found. The PCP-SAFT equation of state is used for thermodynamic modeling. Bubble-point pressures calculated from PCP-SAFT deviate from the experimental bubble-point pressure data by only 0.68 % (2-MF/*iso*-octane) and 0.61 % (DNBE/2-MTHF).

The Cailletet setup and measurement method was continuously refined into a remarkably accurate measurement method over many years. To emphasize only one of many details: for measurement of pressure, instead of relying on the approximate $g = 9.81\,\mathrm{m\,s^{-2}}$, the gravitational force in Delft is explicitly taken into account. However, the remarkable accuracy comes at a price: extensive infrastructure is needed for the very sophisticated procedure. In particular, sample preparation based on the freeze/thaw technique requires handling of liquid nitrogen and mercury. As mercury is employed to seal the samples from atmosphere, the setups contain hot mercury at very high pressures. As a result, strict safety precautions are necessary during measurements. In summary, VLE characterization using the Cailletet setup requires considerable instrumental and experimental effort.

Chapter 4

RAMSPEQU: A milliliter-scale setup for VLE

4.1 Introduction

In the last Chapter, we employed the well-established Cailletet method for VLE characterization. We learned that the remarkable accuracy of the Cailletet method requires extensive infrastructure and considerable effort. Additionally, in Section 2.2, limitations of current VLE measurement methods are discussed. In this Chapter, we aim to reduce the immense effort and necessary infrastructure that is involved with the procedure used in Chapter 3. Furthermore, the limitations discussed in Section 2.2 are tackled by introducing a novel milliliter-scale setup for the efficient characterization of VLE: RAMSPEQU (**Ram**an **S**pectroscopic **P**hase **Equ**ilibrium Characterization). RAMSPEQU aims at easy handling, and substance-saving and rapid VLE data collection.

We determine phase compositions in both phases by Raman spectroscopic measurements. Complete $pTxy$-data sets are collected. By analyzing phase compositions using Raman spectroscopy, errors due to sampling are avoided. The setup is operated at low pressure conditions, which is less demanding for the measurement setup itself, but more challenging for the spectroscopic characterization of vapor phase compositions due to the low density in the vapor phase. We measure Raman spectra using a highly sensitive CCD sensor resulting in short analysis times. An integrated workflow is used that efficiently combines sample preparation, pure component vapor pressure measurement, calibration of Raman signals and phase equilibrium characterization of mixtures. The integrated workflow saves both time and substance. We use in-situ degassing for sample preparation, reducing the consumption of chemicals. Furthermore, in-situ degassing is attractive since it saves time and reduces the necessary infrastructure for sample preparation. We avoid large equilibrium cell volumes: Reliable phase equilibrium measurements can be conducted in a small equilibrium cell of less than 3 ml size as sampling is avoided. The small volume of the equilibrium cell further reduces the amount of chemicals needed for phase equilibrium measurements. In addition, the small volume of the equilibrium cell enhances heat and mass transfer, reducing equilibration times.

The experimental RAMSPEQU setup is introduced in Section 4.2. The integrated workflow is presented in Section 4.3. Details on the spectral analysis and Raman signal calibration are discussed from a theoretical point of view in Section 4.4. In Section 4.5, we present VLE data on pure components (methyl tert-butyl ether (MTBE), *iso*-octane, toluene and ethanol) and a binary mixture (MTBE + *iso*-octane) from RAMSPEQU. Our experimental data is compared to literature data for validation of the setup. We determine deviations quantitatively using the PCP-SAFT EOS. Finally, in Section 4.6 the results are summarized and conclusions are drawn regarding the efficient characterization of VLE using the RAMSPEQU setup

Major parts of this Chapter are reproduced by permission of Elsevier from:

> Liebergesell, B., Flake, C., Brands, T., Koß, H.-J., and Bardow, A. (2017) A milliliter-scale setup for the efficient characterization of isothermal vapor-liquid equilibria using Raman spectroscopy, *Fluid Phase Equilibria*, 446:36-45.
>
> Contribution report: Writing the draft, principal author, conceptual design of the setup and measurement procedure, planning and conducting the experiments, data evaluation.

PCP-SAFT parameter fitting for this Chapter has been performed by S. Kaminski.

4.2 Experimental setup

In this Section, we describe the experimental RAMSPEQU setup. The RAMSPEQU setup is sketched in Figure 4.1. The description starts with the central part of the equilibrium cell, a glass cuvette, and ends with the outer parts of the RAMSPEQU setup.

The glass cuvette contains the sample and has a volume of less than 3 ml. The glass gives optical access to laser-light irradiation, and observation of scattered light from both the liquid phase and the vapor phase. The cuvette is held in place by a hollow screw at the bottom of the surrounding water bath. The center of the screw is

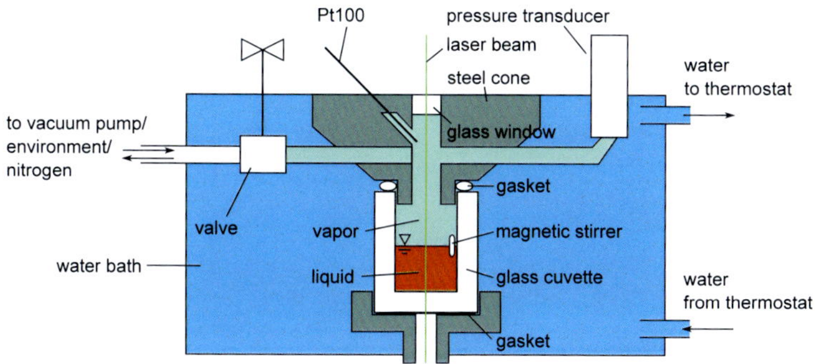

Figure 4.1: VLE setup for the **Raman** **S**pectroscopic **P**hase **Equ**ilibrium Characterization (RAMSPEQU). The equilibrium cell consists of a glass cuvette connected to a steel cone. The cone contains pathways for a Pt100 resistance thermometer, a pressure transducer and a valve. The whole equilibrium cell is surrounded by a water bath for temperature control. The temperature of the water bath is kept constant using a thermostat.

hollow for the laser-light irradiation. The light of a frequency doubled Nd:YAG-laser (Coherent, Verdi G SLM 2, 532 nm, 2 W) passes vertically through liquid and vapor phase contained in the cuvette. Scattered light from the probe volume is collected in 90° angle via the front window. The irradiation laser light exits through a glass window at the top to avoid excessive heating of the sample. The window and a gasket seal the equilibrium cell against the surrounding atmosphere.

The Raman signal is recorded by an optical multichannel analyzer (spectrograph: Princeton Instruments IsoPlane SCT 320, detector: Princeton Instruments ProEM 1600[4] CCD camera). Elastically scattered laser light is blocked by a notch filter (Semrock 532 nm StopLine® single-notch filter) in front of the entrance slit of the spectrograph.

Inside the glass cuvette, a magnetic stirrer accelerates mass transfer between the phases for faster equilibration. The glass cuvette is connected to a stainless steel cone. The glass cuvette and the steel cone form the actual equilibrium cell. All parts of the setup that come into contact with the sample (namely, the equilibrium cell) are immersed into the water bath for temperature control. The temperature is controlled by a thermostat. The connection of the glass cuvette and the steel cone is sealed against the surrounding water bath by a gasket (KALREZ® 6375). Inside the steel cone, the vapor phase is in contact with a Pt100 resistance thermometer (Class

A, 223.15 - 523.15 K, uncertainty 0.2 K) and a pressure transducer (WIKA pressure transducer P-31, 0 - 400 kPa abs., uncertainty 0.2 kPa). In this work, type B uncertainties are reported under the assumption of a rectangular distribution according to NIST recommendation (Chirico et al., 2003). The pressure transducer is immersed into the water bath to avoid condensation of fluid on the membrane of the pressure transducer. Pressure readings are actively temperature compensated. Pressure and temperature are monitored and recorded once per second during equilibration and measurement. The Pt100 can be removed from its mounting to open the equilibrium cell. The equilibrium cell can then be filled with liquid substances by a syringe. The equilibrium cell is connected by a valve to a vacuum pump, the surrounding atmosphere, and the nitrogen supply. The valve is also immersed in the water bath. Via the valve, gaseous substances can be removed from the equilibrium cell, e.g., during degassing, pressure can be equalized to ambient pressure or nitrogen can be fed into the cell.

The setup allows characterization of phase equilibria in a temperature range of 278 K - 338 K and a pressure range of 0.2 kPa - 400 kPa.

All chemicals used in this Chapter and their purities are listed in Table B.1.

4.3 Workflow

In this Chapter, we determine complete VLE data sets: $pTxy$. For this purpose, pressure p, temperature T, liquid phase composition x and vapor phase composition y have to be measured in equilibrium. Pressure p and temperature T are measured by the pressure and temperature sensors (see Section 4.2). The composition of the vapor phase y and liquid phase x are determined based on the Raman signal of the respective phases.

To correlate the Raman signal with composition, a calibration constant k has to be determined. The calibration constant is specific for both, the vapor and the liquid phase (Schrötter and Klöckner, 1979). To reduce the necessary amount of time and substances, we use an integrated workflow for the calibration procedure and the actual measurement of the phase equilibria. The workflow is described in Sections 4.3.1 to 4.3.4. A flowchart illustrating the workflow is given in Figure 4.2.

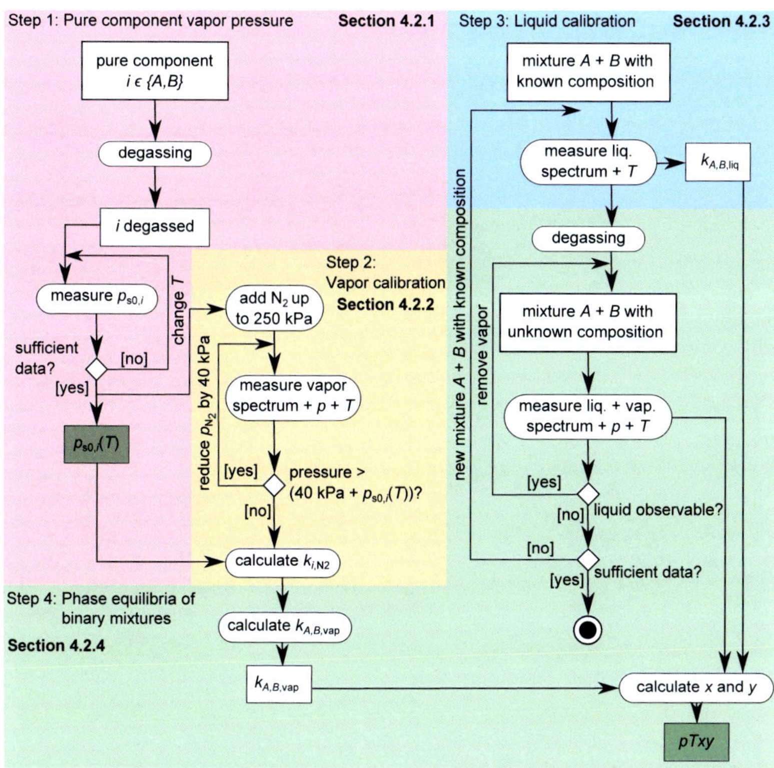

Figure 4.2: Flowchart illustrating the workflow for the characterization of the binary VLE of species A and B using the RAMSPEQU setup. In Step 1, pure component vapor pressures $p_{s0,i}(T)$ are measured (Section 4.3.1). In Step 2, the pure component is used for calibration of the vapor phase (Section 4.3.2). In Step 3, the liquid phase is calibrated (Section 4.3.3). In Step 4, the binary mixture is used for the characterization of phase equilibria (Section 4.3.4). The calibration described in Sections 4.3.2 and 4.3.3 is conducted at the same temperature as the phase equilibrium measurements (Section 4.3.4). Calculation of the calibration constants k is explained in Sections 4.4.1 and 4.4.2. The main results $p_{s0,i}(T)$ and $pTxy$ are highlighted by green boxes.

4.3.1 Step 1: Vapor pressure measurement of pure components

Prior to any experimental measurement in the RAMSPEQU setup, the equilibrium cell is thoroughly cleaned using ethanol as the cleaning solvent. To remove solvent residue after cleaning, the equilibrium cell is evacuated using the vacuum pump.

Thereafter, the cell is filled with nitrogen gas until a pressure slightly above ambient pressure is reached. The cell is filled with nitrogen to avoid contamination of the equilibrium cell with, e.g., atmospheric humidity, upon opening the equilibrium cell. The equilibrium cell is opened by removing the Pt100 from its mounting. The equilibrium cell is filled with a liquid pure component by a syringe. Subsequently, we reattach the Pt100 to close the equilibrium cell. The pure component is degassed in-situ by opening the equilibrium cell's valve to vacuum repeatedly. Degassing is completed once the pressure reading reached in equilibrium remains the same after two consecutive degassing cycles. After degassing is completed, the vapor pressure of the pure component is measured. For the vapor pressure measurement, the desired temperature is adjusted via the thermostat. Thereafter, the phases are equilibrated. Throughout this work, we assume that equilibrium is reached as soon as the temperature measured in the cell is constant within 0.1 K and the pressure is constant within 0.2 kPa for at least 5 minutes. After recording a vapor pressure, temperature is changed via the thermostat to measure the next vapor pressure. The workflow is repeated until sufficient data is generated. In this work, we typically collect 6 to 13 vapor pressure data points using the RAMSPEQU setup.

4.3.2 Step 2: Vapor phase calibration

The composition of the vapor phase y is determined based on the Raman signal of the vapor phase. To determine an unknown composition of a mixture of components A and B in the vapor phase from its Raman signal, a calibration constant $k_{A,B,\text{vap}}$ has to be determined (relative calibration, see Section 2.2). For this calibration, the Raman spectra of mixtures of known composition are collected. In this Section, the calibration procedure is described for a binary mixture. To study a mixture with more than two components, the procedure can be directly extended.

In a straightforward calibration procedure, vapor mixtures of species A and B of known composition would be prepared and their Raman signal would be measured. The calibration constant $k_{A,B,\text{vap}}$ could be determined directly from the collected Raman signals and their respective known compositions (as is done for the liquid phase, see Sections 4.3.3 and 4.4.2). However, such a straightforward calibration procedure is difficult for the vapor phase: First, vapor mixtures with well-known composition have to be prepared. Thereafter, the mixtures have to be added into a fully evacuated equilibrium cell. Additionally, even partial condensation of the mixtures has to be strictly avoided to exclude any change of the composition in the vapor phase.

To avoid the difficult calibration procedure in the vapor phase, we propose an indirect calibration procedure using nitrogen as non-condensable gas. For the indirect calibration, we prepare a mixture of species A and nitrogen of known composition; the composition is known from the behavior of ideal gas mixtures (see Equation 4.1 and 4.2). The experimental workflow for the preparation of the mixtures is described below. For calibration, the compositions of the mixtures are correlated with their Raman signals to determine a calibration constant $k_{A,\mathrm{N_2},\mathrm{gas}}$ (see Section 4.4.1). The same procedure is conducted for species B to determine a calibration constant $k_{B,\mathrm{N_2},\mathrm{gas}}$. Based on the calibration constants $k_{A,\mathrm{N_2},\mathrm{gas}}$ and $k_{B,\mathrm{N_2},\mathrm{gas}}$ we can calculate the desired calibration constant $k_{A,B,\mathrm{vap}}$ (see Section 4.4.1). The calibration constant $k_{A,B,\mathrm{vap}}$ enables us to calculate the unknown composition in a mixture of A in B in the vapor phase from its Raman signal.

The preparation of a mixture of species i (A respectively B) and nitrogen in the gaseous (gaseous: both gas and vapor) phase starts after the vapor pressure of the pure component i ($p_{\mathrm{s0},i}$) has been measured as described in Section 4.3.1. Subsequently, gaseous nitrogen is fed into the cell (e.g., until a pressure of about 250 kPa is reached). The overall pressure p_{exp} is measured in equilibrium. Raman spectra of the gaseous phase are recorded (typical exposure time: 10 seconds).

For the indirect calibration method, two assumptions are made: (1) The vapor pressure $p_{\mathrm{s0},i}$ of component i is not influenced by the addition of gaseous nitrogen. (2) Due to low pressures, the gaseous phase can be treated as an ideal gas. With

$$p_{\mathrm{exp}} = p_{\mathrm{s0},i} + p_{\mathrm{N_2}} \tag{4.1}$$

and

$$\frac{p_{\mathrm{s0},i}}{p_{\mathrm{s0},i} + p_{\mathrm{N_2}}} = \frac{n_i}{n_i + n_{\mathrm{N_2}}}, \tag{4.2}$$

where n_i denotes the amount of substance of species i in the gas phase. As $p_{\mathrm{s0},i}$ (vapor pressure of component i) and p_{exp} (overall pressure measured in equilibrium) are known, the partial pressure of nitrogen $p_{\mathrm{N_2}}$ can be determined from Equation 4.1. Consequently, the composition of the mixture in the gaseous phase is known (Equation

4.2).

The composition of the gaseous phase is changed by partly depressurizing, e.g., reducing p_{exp} in steps of 40 kPa towards the pure component vapor pressure $p_{\mathrm{s0},i}$. Starting at measuring the overall pressure p_{exp}, the workflow described in this Section is repeated to collect further spectra of mixtures of known composition. In this work, we typically collect 5 spectra of mixtures of known composition for each component i.

4.3.3 Step 3: Liquid phase calibration

To determine an unknown composition in the liquid phase from its Raman signal, another calibration constant $k_{A,B,\mathrm{liq}}$ has to be determined for the relative calibration method. For this purpose, spectra of liquid mixtures of known composition are measured. The actual calculation of the calibration constant is discussed in Section 4.4.1.

The liquid phase calibration starts with preparation of the cell by the same workflow as described in Section 4.3.1 for pure component vapor pressure measurements up to and including filling the cell with nitrogen gas. The equilibrium cell is opened by removing the Pt100 from its mounting. The equilibrium cell is filled with a liquid mixture of known composition by a syringe. Subsequently, we reattach the Pt100 to close the equilibrium cell. As soon as thermal equilibrium is reached, a Raman spectrum of the liquid phase of known composition is measured for calibration (typically 1 second exposure time). We assume that evaporation within the equilibrium cell has no significant influence on the liquid phase composition. We base this assumption on the small overall cell volume (< 8 ml) and the low molar holdup in the vapor phase at the low pressures studied in this work.

Subsequently, the workflow described in the next Section 4.3.4 is conducted to analyze several phase equilibria from the calibration sample. Thereafter, the workflow described in the present Section 4.3.3 is repeated to collect further spectra of mixtures of known compositions. In this Chapter, we use 5 independent loadings of the equilibrium cell for the calibration of the liquid phase.

4.3.4 Step 4: Characterization of phase equilibria

Phase equilibrium measurements are conducted with the same loading of the cell that has previously been used for the calibration of the liquid phase. After the calibration described in Section 4.3.3 is completed, the mixture is degassed by opening the equilibrium cell's valve to vacuum repeatedly. During degassing, the composition of the liquid mixture changes, as the components that are enriched in the vapor phase are preferentially removed from the equilibrium cell. Change in composition during degassing is not an issue as the composition of vapor and liquid phase are analyzed during the phase equilibrium characterization. After degassing, the phases are equilibrated. In equilibrium, we measure Raman spectra of the liquid (typically 1 second exposure time) and of the vapor phase (typically 10 seconds exposure time).

For the next $pTxy$-data set, we change the overall composition by opening the equilibrium cell's valve to vacuum again. After changing the composition, the system is once again equilibrated and Raman spectra of liquid and vapor phase are measured. Consequently, with a single loading of the equilibrium cell, several $pTxy$-data sets can be collected. The procedure can be repeated as long as signal from the liquid phase can be measured. From a single loading of the equilibrium cell, we measure up to 5 $pTxy$-data sets.

We repeat the steps described in Sections 4.3.3 and 4.3.4 to collect several calibration spectra of mixtures of known compositions and to analyze the phase equilibria.

4.4 Spectral Analysis and Signal Calibration

The intensity of a Raman signal emitted by a component is directly proportional to the number of molecules of the component that are excited (Pelletier, 2003). The Raman spectrum of a mixture contains the Raman signal of all components. Consequently, from the Raman spectrum of a mixture, we can determine the composition of the mixture.

In this work, we discriminate the Raman signals of the individual components within a mixture Raman spectrum using indirect hard modeling (IHM) (Alsmeyer et al., 2004). IHM models the Raman spectrum of each component as a sum of peak-shaped Voigt functions. Mixtures are modeled as weighted sum of pure component models. In mixture spectra, nonlinear changes (such as peak shifts) can occur due to molecular interactions. In IHM, nonlinear changes in mixture spectra can be considered by varying the parameters of the Voigt functions. Since IHM closely follows the physical principles leading to mixture spectra, only few mixture spectra at known composition are needed for calibration (Alsmeyer et al., 2004).

The major challenge in the analysis of low-pressure VLE is the determination of the vapor phase composition. In low-pressure VLE, the number of molecules in the vapor phase is small. The small number of molecules in the vapor phase leads to a low intensity of the Raman signal. Due to the low intensity Raman signal of the vapor phase, changing background signals (e.g., from the surrounding water bath) can be significant. In this Chapter, we therefore analyze the vapor phase Raman spectra using first-derivative indirect hard modeling (FD-IHM) (Beumers et al., 2018): Analysis of the first derivative of a spectrum is very robust with respect to changing background signals. In FD-IHM, Raman spectra are modeled in the same way as in classical IHM, but the whole analysis is based on the first derivative of the spectrum.

Whenever spectra are recorded throughout this work, we record at least three spectra to eliminate possible cosmic ray events. Cosmic ray events are eliminated by calculating the temporal pixel-by-pixel median of at least three spectra for each measurement (Croke, 1995).

Exemplary Raman spectra of two pure components (*iso*-octane and MTBE) and spectra of their mixture are shown in Figure 4.3. Vapor phase spectra show no significant noise (Figure 4.3a) but fewer details can be resolved compared to the liquid phase (Figure 4.3b).

4.4.1 Calibration for vapor phase analysis

In this Section, the indirect calibration procedure for vapor phase analysis is explained. The corresponding experimental workflow is discussed in Section 4.3.2.

The amount of substance n_i, of any species i, is proportional to the detected Raman signal S_i:

$$n_i = k_i \cdot S_i \tag{4.3}$$

with k_i as constant of proportionality. In IHM, the Raman signal S_i is the area under the Voigt peak functions corresponding to one component in the fitted mixture spectrum.

From Equation 4.3, we get for a mixture of species i and N_2 in the gaseous phase

$$\frac{n_i}{n_i + n_{\mathrm{N}_2}} = \frac{k_i \cdot S_i}{k_i \cdot S_i + k_{\mathrm{N}_2} \cdot S_{\mathrm{N}_2}} = \frac{k_{i,\mathrm{N}_2} \cdot S_i}{k_{i,\mathrm{N}_2} \cdot S_i + S_{\mathrm{N}_2}} \tag{4.4}$$

$$\text{with } k_{i,\mathrm{N}_2} = \frac{k_i}{k_{\mathrm{N}_2}}$$
$$\text{and } i \in \{A, B\}.$$

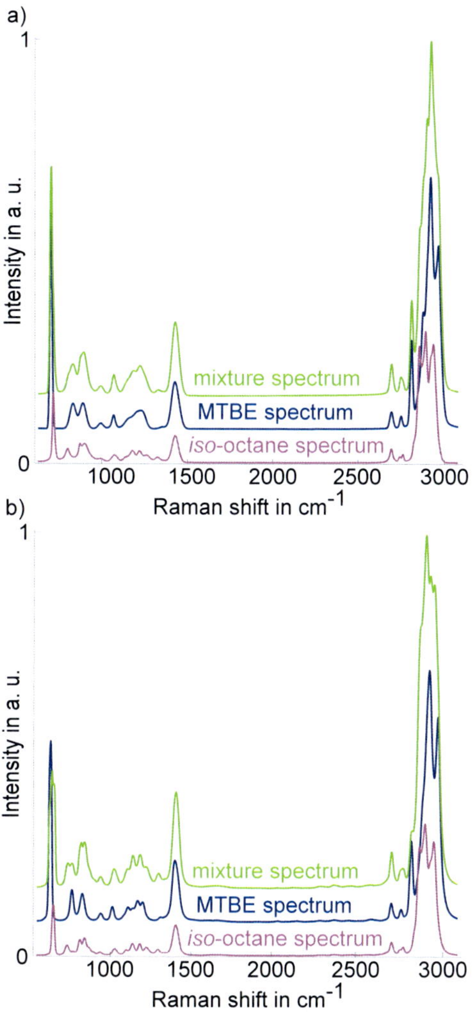

Figure 4.3: a) Vapor phase Raman spectra of the pure components *iso*-octane (magenta) and MTBE (blue), as well as their mixture spectrum (green, $y_{\mathrm{MTBE}} = 0.736$). b) Liquid phase Raman spectra of the pure component *iso*-octane (magenta) and MTBE (blue), as well as their mixture spectrum (green, $x_{\mathrm{MTBE}} = 0.410$). Spectra were recorded at a temperature of $T = 318.1\,\mathrm{K}$ and the respective equilibrium pressure.

The calibration constant k_{i,N_2} is determined by fitting it to varied ratios of $n_i/(n_i+n_{\mathrm{N}_2})$ and the corresponding Raman signals S_i. Eventually, we want to determine the mole fraction y_A of species A (and y_B of species B) in a mixture of the species A and B. Similar to Equation 4.4 follows:

$$\begin{aligned} y_A &= \frac{n_A}{n_A + n_B} = \frac{k_A \cdot S_A}{k_A \cdot S_A + k_B \cdot S_B} = \frac{\frac{k_A}{k_{\mathrm{N}_2}} \cdot S_A}{\frac{k_A}{k_{\mathrm{N}_2}} \cdot S_A + \frac{k_B}{k_{\mathrm{N}_2}} \cdot S_B} \\ &= \frac{k_{A,\mathrm{N}_2} \cdot S_A}{k_{A,\mathrm{N}_2} \cdot S_A + k_{B,\mathrm{N}_2} \cdot S_B} = \frac{k_{A,B} \cdot S_A}{k_{A,B} \cdot S_A + S_B} \end{aligned} \tag{4.5}$$

$$\text{with } k_{A,B} = \frac{k_{A,\mathrm{N}_2}}{k_{B,\mathrm{N}_2}}.$$

Consequently, with known k_{A,N_2} and k_{B,N_2}, the mole fraction y_A of species A in a mixture of the species A and B can be determined based on the spectroscopically observed Raman signals S_A and S_B in the vapor phase.

To study a mixture with more than two components, the procedure is readily extended. With n_C the number of components, Equation 4.5 becomes:

$$y_A = \frac{n_A}{\sum_{i=1}^{n_C} n_i} = \frac{k_{A,\mathrm{N}_2} \cdot S_A}{\sum_{i=1}^{n_C} k_{i,\mathrm{N}_2} \cdot S_i} \tag{4.6}$$

$$\text{with } k_{i,\mathrm{N}_2} = \frac{k_i}{k_{\mathrm{N}_2}}.$$

4.4.2 Calibration for liquid phase analysis

In this Section, the calibration procedure for the liquid phase is explained. This calibration is conducted directly by measuring spectra of mixtures with known compositions and correlating the spectra with the compositions. The corresponding experimental workflow is described in Section 4.3.3. The calibration constant $k_{A,B,\mathrm{liq}}$ is required to calculate the mole fraction x_A of species A (and x_B of species B) in a liquid mixture of unknown composition (see Equation 4.7).

Equation 4.3 is also valid in the liquid phase. The calibration constant $k_{A,B,\mathrm{liq}}$ can directly be determined from Equation 4.7 by fitting the calibration constant $k_{A,B,\mathrm{liq}}$

to varied ratios of $n_A/(n_A + n_B)$ and their corresponding Raman signals S_i.

$$x_A = \frac{n_A}{n_A + n_B} = \frac{k_{A,\mathrm{liq}} \cdot S_A}{k_{A,\mathrm{liq}} \cdot S_A + k_{B,\mathrm{liq}} \cdot S_B} = \frac{k_{A,B,\mathrm{liq}} \cdot S_A}{k_{A,B,\mathrm{liq}} \cdot S_A + S_B} \tag{4.7}$$

To study a mixture with more than two components, e.g., four components as in Chapter 5, the procedure is readily extended (see Equation 4.6) by adding more components to the mole fraction that is the starting point in Equation 4.7.

4.5 Validation: VLE experiments and results

Pure component vapor pressures of four components and binary VLE data of the test system MTBE + *iso*-octane are measured and compared to literature data for validation of the novel RAMSPEQU setup. The binary system for validation is chosen as it is recommended based on consistency tests in literature (Goral et al., 2003).

4.5.1 Equilibration

Equilibration is fast in the RAMSPEQU setup. Equilibration starts with the rise in temperature and pressure after closing the equilibrium cell's valve to vacuum (Figure 4.4, minute 0). Pressure and temperature can already be considered constant after 10 minutes (Figure 4.4). The graph also shows that leaking can be neglected. Equilibration is usually achieved within 15 minutes in this work.

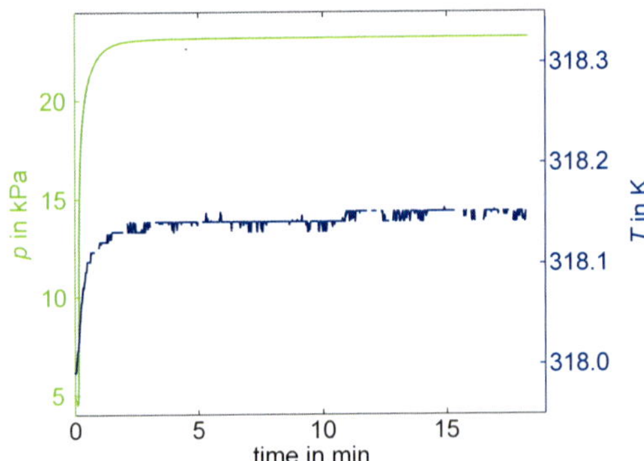

Figure 4.4: Equilibration process for ethanol at 318.2 K. Pressure p and temperature T are plotted over time. The scatter in temperature is smaller than the measurement uncertainty ($u(T) = 0.2\,\mathrm{K}$).

4.5.2 Pure component vapor pressures

Pure component vapor pressures from literature are reproduced using the RAMSPEQU setup (Figure 4.5). The vapor pressures of MTBE, ethanol, *iso*-octane and toluene are measured at temperatures of 278 - 338 K and pressures of up to 137 kPa kPa (Table B.3 and Figure 4.5).

The pure component vapor pressures are in agreement with literature data within the measurement uncertainties ($u(T) = 0.2\,\mathrm{K}$, $u(p) = 0.2\,\mathrm{kPa}$). Thus, in-situ degassing is found to be sufficient for accurate phase equilibrium characterization. Complete degassing of the components used in this work is usually achieved within less than 45 minutes.

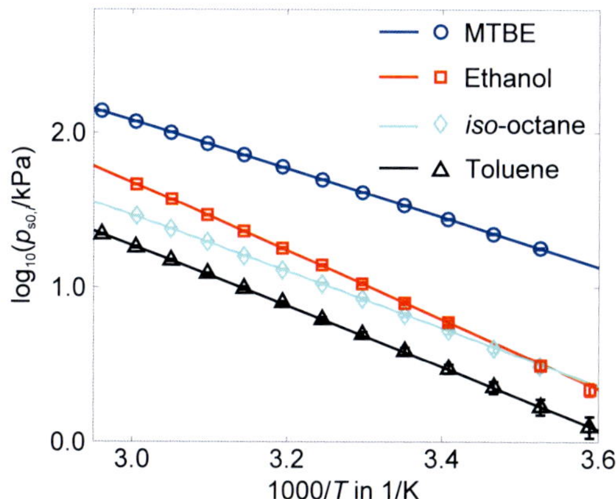

Figure 4.5: Comparison of experimentally determined vapor pressure $p_{s0,i}$ of pure substances used in this work as a function of temperature T, measured with the RAMSPEQU setup (symbols), to vapor pressure data from literature (lines). Lines are calculated from the Antoine equation for MTBE (Boublík et al., 1985), *iso*-octane (Horstmann et al., 1999), and toluene (Negadi et al., 2009). For ethanol (EtOH), the vapor pressure line is calculated from an ancillary function to a fundamental equation for the calculation of the thermodynamic properties of ethanol (Dillon and Penoncello, 2004). Error bars indicate the instrumental measurement uncertainty and are only visible at low pressure. The measurement uncertainties are $u(T) = 0.2\,\mathrm{K}$ and $u(p) = 0.2\,\mathrm{kPa}$.

4.5.3 Binary system

We validate the RAMSPEQU setup by comparing our experimental data (Table B.4) to binary VLE data for the system MTBE + *iso*-octane from literature (Bernatová and Wichterle, 2001) (Figure 4.6).

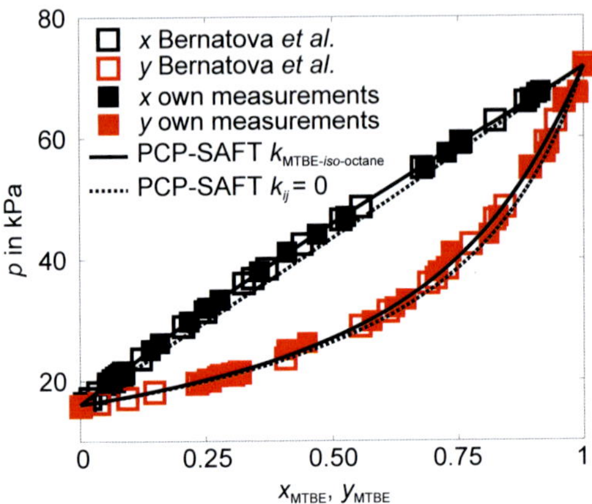

Figure 4.6: Isothermal pressure - liquid phase composition - vapor phase composition (p-x-y)-diagram representing the vapor-liquid equilibrium of MTBE + *iso*-octane at $T = 318.1\,\mathrm{K}$. Comparison of experimental results determined using RAMSPEQU to results reported in literature (Bernatová and Wichterle, 2001). Solid lines represent phase equilibrium data calculated from PCP-SAFT based on the parameters shown in Table 4.1 and the binary interaction parameter $k_{\text{MTBE-}iso\text{-octane}}$. For comparison, we show the phase behavior predicted by PCP-SAFT based on pure component parameters only ($k_{ij} = 0$, dashed lines). The measurement uncertainties are $u(T) = 0.2\,\mathrm{K}$, $u(p) = 0.2\,\mathrm{kPa}$, $u(x) = 0.008$ and $u(y) = 0.01$.

The uncertainty of composition of the liquid phase is $u(x) = 0.008$. The reported uncertainty of the composition of the liquid phase $u(x)$ is the root mean squared error of cross-validation (RMSECV) from calibration where we used leave-one-out cross-validation. The RMSECV allows to estimate the uncertainty of the calibration when it is applied to unknown data. The RMSECV found is independent of composition. Based on the calibration data, we could not detect a significant dependence of uncertainty on composition. The RMSECV of the vapor phase calibration is $\text{RMSECV}(y) = 0.002$. However, the uncertainty of the composition of the vapor

Table 4.1: Pure component PCP-SAFT parameters. Parameters (m_i, σ_i, ϵ_i) are adjusted to one liquid density (Table B.2) and the vapor pressure values shown in Figure 4.5 and Table B.3. Dipole and quadrupole parameters (μ_i, θ_i) are calculated as described in detail by van Nhu et al. (2008). k_B is the Boltzmann constant.

Component	m	σ/Å	(ϵ/k_B) / K	μ/D	θ/DÅ
MTBE	2.9758	3.6931	231.302	1.278	3.862
iso-octane	3.3781	3.9832	239.578	0.128	0.711

phase is larger than the RMSECV(y) from calibration due to a changing laser irradiation in the vapor phase over the course of an experiment. These changes are due to the changing position of the phase boundary relative to the magnetic stirrer. We therefore report a higher uncertainty of $u(y) = 0.01$ based on the comparison of the experimental results to phase equilibrium data from PCP-SAFT (see below).

Our experimental data is modeled using the PCP-SAFT EOS. Pure component PCP-SAFT parameters are fitted based on one liquid density (Table B.2) and all pure component vapor pressure data (Table B.3) obtained in this Chapter. The resulting pure component PCP-SAFT parameters are listed in Table 4.1. The binary interaction parameter for the dispersion interactions k_{ij} is fitted to all $pTxy$-data sets obtained in this Chapter (Table B.4). The resulting binary interaction parameter is $k_{\text{MTBE-}iso\text{-octane}} = 8.262 \cdot 10^{-3}$.

As shown in Figure 4.6, the model agrees well with our experimental data. To evaluate the quality of fit, we compare our data to phase equilibrium data from PCP-SAFT. We evaluate the quality of fit as follows: We determine the absolute average deviation in terms of pressure Δp_{fit} (see Equation 4.8) and vapor phase composition Δy_{fit} (see Equation 4.9) of experimentally determined data from the calculated phase equilibrium data for a given temperature and liquid phase composition:

$$\Delta p = \frac{1}{N}\sum_{i=1}^{N} \Delta p_i = \frac{1}{N}\sum_{i=1}^{N} \left| \frac{p_i^{\text{exp}} - p_i^{\text{cal}}}{p_i^{\text{exp}}} \right| \tag{4.8}$$

$$\Delta y = \frac{1}{N}\sum_{i=1}^{N} \Delta y_i = \frac{1}{N}\sum_{i=1}^{N} \left| \frac{y_i^{\text{exp}} - y_i^{\text{cal}}}{y_i^{\text{exp}}} \right| \tag{4.9}$$

with N, the number of experimentally determined property values, the superscript exp indicates experimental data, and the superscript cal indicates data calculated from PCP-SAFT.

We find deviations of $\Delta p_{\text{fit}} = 0.5\,\%$ and $\Delta y_{\text{fit}} = 1.0\,\%$, indicating a good quality of fit.

Additionally, we use the phase equilibrium calculations based on PCP-SAFT to compare our data to literature data. Again, we determine the deviations in terms of pressure $\Delta p_{\text{literature}}$ (see Equation 4.8) and vapor phase composition $\Delta y_{\text{literature}}$ (see Equation 4.9). However, instead of our own experimental data, we compare the experimental literature data to the calculated phase equilibrium data for a given temperature and liquid phase composition. We find deviations of $\Delta p_{\text{literature}} = 0.8\,\%$ and $\Delta y_{\text{literature}} = 0.9\,\%$, indicating a very good agreement of the experimental data from literature with the phase equilibrium calculations from PCP-SAFT based on experimental data from our RAMSPEQU setup.

If the binary interaction parameter k_{ij} is not fitted but set to zero, the prediction of the mixture behavior from pure component data by PCP-SAFT is still qualitatively in very good agreement with the measured data (see Figure 4.6). If we quantify the deviation of the calculated phase equilibrium data to our own experimental data ($\Delta p_{\text{fit},k_{ij}=0}$, $\Delta y_{\text{fit},k_{ij}=0}$) and the experimental data from literature ($\Delta p_{\text{literature},k_{ij}=0}$, $\Delta y_{\text{literature},k_{ij}=0}$), we find deviations of $\Delta p_{\text{fit},k_{ij}=0} = 3.7\,\%$ and $\Delta y_{\text{fit},k_{ij}=0} = 3.7\,\%$ respectively $\Delta p_{\text{literature},k_{ij}=0} = 3.2\,\%$ and $\Delta y_{\text{literature},k_{ij}=0} = 5.1\,\%$. The resulting deviations show that even though the value of the fitted binary interaction parameter $k_{\text{MTBE-}\textit{iso}\text{-octane}} = 8.262 \cdot 10^{-3}$ is rather small, the resulting phase equilibrium calculations already differ significantly from the purely predictive binary phase equilibrium calculations.

4.6 Conclusion

The **Ram**an **S**pectroscopic **P**hase **Equ**ilibrium Characterization (RAMSPEQU)-setup is presented for the non-invasive characterization of isothermal low-pressure VLE. The RAMSPEQU setup and measurement procedure is validated through comparison to literature data for (1) vapor pressures of the pure components MTBE, ethanol, *iso*-octane and toluene and for (2) the binary VLE of MTBE and *iso*-octane.

In-situ degassing is found to be sufficient for accurate phase equilibrium characterization. Using in-situ degassing, we reduce the consumption of chemicals and accelerate the degassing procedure. Additionally, in-situ degassing enables easy handling of the setup: Complete evacuation of the equilibrium cell prior to filling it is not necessary. Liquid substances can easily be added into the equilibrium cell at atmospheric conditions. Lengthy freeze/thaw cycles and the necessary infrastructure are avoided.

The novel RAMSPEQU setup avoids large cell volumes: Less than 3 ml of substance per loading of the equilibrium cell are sufficient for phase equilibrium characterization. Consequently, novel substances, that are often expensive or even available only in small amounts, can be studied at reduced cost. Furthermore, harmful substances or novel substances that have to be considered harmful until proven otherwise, can be studied at reduced risk. Additionally, the small volume enhances heat and mass transfer resulting in reduced equilibration times.

To further reduce the amount of time and substances necessary, we use an integrated workflow that enables a liquid phase calibration measurement and the measurement of several $pTxy$-data sets from a single filling of the equilibrium cell. The procedure that is used for degassing is also employed to change the overall composition of mixtures inside the equilibrium cell. Typically, up to 5 $pTxy$-data sets can be measured with one loading of the equilibrium cell.

Non-invasive Raman spectroscopic analysis is used to determine the composition of the vapor and liquid phase. Nonlinear spectral changes due to molecular interactions are accounted for as indirect hard modeling (IHM) is employed for spectral analysis. The composition of the vapor phase agrees with values from the literature, even at the challenging low pressure conditions. The Raman signal is measured using a highly sensitive CCD sensor for light detection. Consequently, analysis times are reduced to several seconds. As phase compositions are determined non-invasively, sampling is avoided and thus a major error source for analytical phase equilibrium data.

To enable composition measurements in the vapor phase, we use an indirect calibration procedure by studying calibration mixtures with nitrogen. Thereby, we avoid the cumbersome and error-prone preparation of vapor samples. However, the preparation of calibration mixtures is based on the assumption of ideal gas behavior in the gaseous phase. Therefore, the indirect calibration procedure is limited to low pressures.

The combination of the accelerated degassing procedure and the reduced time for equilibration and analysis leads to rapid data generation: up to 15 $pTxy$-data sets can be measured per workday.

RAMSPEQU is extendable to all Raman-active and transparent substances. Spectroscopic techniques become particularly advantageous for the characterization of phase equilibria with a larger number of components.

Chapter 5

VLE of a quaternary model biofuel using RAMSPEQU

5.1 Introduction

VLE data of multicomponent mixtures is the most commonly required physical property data in the chemical industry (Hendriks et al., 2010). Thus, there is a strong need for VLE data of systems with more than two components. However, a "reasonably complete characterization" of VLE with more than two components "rapidly grows to impractical proportions" (Raal and Mühlbauer, 1998). In the last Chapter, we introduced the fast and substance-saving RAMSPEQU method for VLE characterization. The efficient RAMSPEQU method seems particularly suited for the characterization of multicomponent VLE.

In this Chapter, we characterize the VLE of a quaternary model fuel and its binary subsystems using the RAMSPEQU setup. The model fuel is a blend of two bio-based components, 2-butanone and tetrahydrofuran (THF), with two fossil components, n-heptane and cyclohexane. 2-butanone has recently been identified as a promising biofuel candidate due to favorable combustion characteristics (Hoppe et al., 2016). THF is the simplest hydrofuran and is used here to represent the hydrofuran family. As already introduced in Chapter 3, several furans and hydrofurans have been identified as promising biofuel candidates: 2-methylfuran was shown to reduce hydrocarbon emissions and increase efficiency in comparison to model gasolines (Thewes et al., 2011); 2-methyltetrahydrofuran was shown to reduce or even eliminate soot formation in certain blends (Janssen et al., 2011). THF itself has also received attention as model fuel, e.g., in combustion kinetics (de Bruycker et al., 2017).

To represent linear fossil hydrocarbons in fuels, we chose n-heptane. n-Heptane is often used as model fuel for several combustion modes, e.g., jet fuel, homogeneous charge compression ignition (HCCI) fuel, and diesel fuel (Aichlmayr et al., 2003; Montgomery et al., 2002). To represent cyclic fossil hydrocarbons in fuels, we chose cyclohexane which is also often used as a model fuel (Ra and Reitz, 2011; Klein-Douwel et al., 2009; Peukert et al., 2011).

The RAMSPEQU setup is employed to determine isothermal VLE data for 2-butanone + n-heptane + THF + cyclohexane and its binary subsystems. To further reduce the experimental effort, we determine pTx-data. For thermodynamic modeling, we use the PCP-SAFT EOS (see Section 2.1.1). Experimental and modeling results are discussed in Sections 5.2 (Density measurements), 5.3 (Pure component vapor pressures), 5.4 (Binary systems), and 5.5 (Quaternary systems). The Chapter is summarized and conclusions are drawn in Section 5.6.

Major parts of this Chapter are reproduced by permission of Elsevier from:

> Liebergesell, B., Brands, T., Koß, H.-J., and Bardow, A. (2018) Quaternary isothermal vapor-liquid equilibrium of the model biofuel 2-butanone + n-heptane + tetrahydrofuran + cyclohexane using Raman spectroscopic characterization, *Fluid Phase Equilibria*, 472:107-116.
>
> Contribution report: Writing the draft, principal author, planning and conducting the experiments, modeling of phase equilibria, data evaluation.

5.2 Density measurements

All chemicals used in this Chapter and their purities are listed in Table C.1.

We determine liquid densities at a temperature of $T = 298.15\,\mathrm{K}$ and a pressure of $p = 0.1\,\mathrm{MPa}$ according to the oscillating U-tube principle as described in Section 3.2. Density measurements are solely performed for the estimation of PCP-SAFT parameters (Table 5.2). The experimental liquid density data are listed in Table 5.1. The densities measured in this work agree well with data from literature (Alonso et al., 2010; Azizian and Bashavard, 2008; Mohsen-Nia et al., 2009) (see Table 5.1).

Table 5.1: Experimentally determined pure component densities (ρ) at $p = 0.1\,\text{MPa}$ and $T = 298.15\,\text{K}$ with the standard uncertainties $u(i)$ and density values from the literature.

substance	$\rho/\text{g cm}^{-3}$	
	Experimental	Literature
2-butanone	0.80203	0.799861 (Alonso et al., 2010)
n-heptane	0.68148	0.6819 (Azizian and Bashavard, 2008)
THF	0.88499	0.8829 (Mohsen-Nia et al., 2009)
cyclohexane	0.77611	0.7739 (Azizian and Bashavard, 2008)

$u(p) = 0.001\,\text{MPa}, u(T) = 0.01\,\text{K}, u(\rho) = 0.000\,01\,\text{g cm}^{-3}$

5.3 Pure component vapor pressures

Pure component vapor pressures are measured using the RAMSPEQU setup. The vapor pressures of 2-butanone, n-heptane, THF and cyclohexane are measured at temperatures of $T = 283$ - $333\,\text{K}$ and pressures of $p = 2.8$ - $81.9\,\text{kPa}$ (see Figure 5.1 and Table C.2). The pure component vapor pressures are in agreement with literature data (Forero G. and Velásquez J., 2011) within the measurement uncertainties ($u(T) = 0.2\,\text{K}$, $u(p) = 0.2\,\text{kPa}$, see Figure 5.1).

Our experimental data is modeled using the PCP-SAFT EOS. We fit each set of pure component parameters based on one liquid density (Table 5.1) and 6 vapor pressure data points (Table C.2). The resulting pure component parameters are listed in Table 5.2. We use the PCP-SAFT parameters for modeling in Sections 5.4 and 5.5.

Table 5.2: Pure component PCP-SAFT parameters. Parameters (m_i, σ_i, ϵ_i) were adjusted to one liquid density (Table 5.1) and the vapor pressure values shown in Figure 5.1 and Table C.2. Dipole and quadrupole parameters (μ_i, θ_i) were calculated from quantum mechanics as described in detail by van Nhu et al. (2008). k_B is the Boltzmann constant.

Component	m	$\sigma/\text{Å}$	(ϵ/k_B) / K	μ/D	$\theta/\text{DÅ}$
2-butanone	2.5370	3.6276	252.294	2.974	6.847
n-heptane	3.5868	3.7486	234.709	0.090	2.213
THF	2.2810	3.6170	280.055	1.870	4.941
cyclohexane	2.6045	3.7884	273.833	0.000	0.726

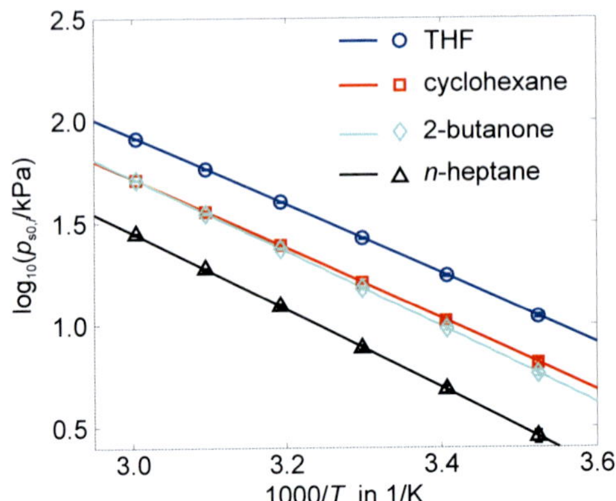

Figure 5.1: Vapor pressure $p_{s0,i}$ of pure components as function of temperature T, measured in this work (symbols), and from literature (lines). Lines are calculated from the Wagner equation for THF, cyclohexane, 2-butanone and n-heptane (Forero G. and Velásquez J., 2011). The measurement uncertainties are $u(T) = 0.2\,\text{K}$ and $u(p) = 0.2\,\text{kPa}$. Error bars indicate the measurement uncertainty and are only visible at low pressure.

5.4 Binary Systems

Isothermal binary vapor-liquid equilibria are characterized using the RAMSPEQU setup. In this Chapter, we determine pTx-data sets only. Consequently, the work-flow described in Section 4.3 is reduced by the steps that are only necessary for the characterization of the vapor phase composition y.

The experimental VLE data of the binary systems 2-butanone + n-heptane, 2-butanone + THF, 2-butanone + cyclohexane, n-heptane + THF, n-heptane + cyclohexane and THF + cyclohexane at $T = 303.2\,\text{K}$ are listed in Table C.3 and shown in Figures 5.2, 5.3 and 5.4. The presented binary phase equilibria are measured using an average of 22 ml of the respective binary mixture. The reported uncertainty in the composition $u(x_{i/j})$ is the RMSECV from calibration where we use leave-one-out cross validation (see Section 4.5.3). Based on calibration data, we could not detect a dependence of $u(x_{i/j})$ on composition.

For thermodynamic modeling, we use the PCP-SAFT EOS. Pure component pa-

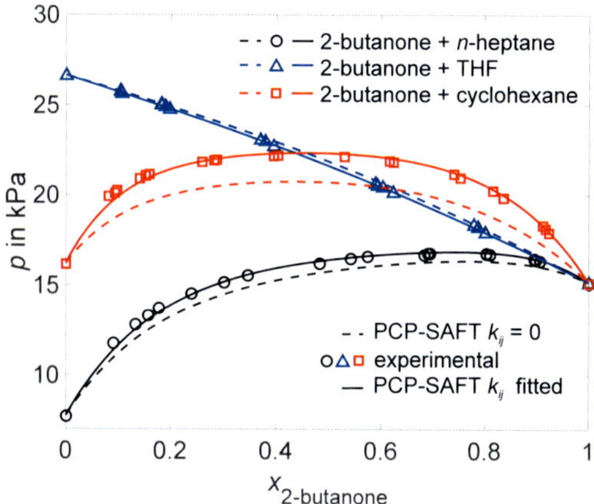

Figure 5.2: Isothermal pressure - liquid-phase composition ($p-x$)-diagram representing the vapor-liquid equilibrium of 2-butanone (1) + n-heptane (2), 2-butanone + THF (3) and 2-butanone + cyclohexane (4) at $T = 303.2\,\text{K}$. The measurement uncertainties are $u(T) = 0.2\,\text{K}$, $u(p) = 0.2\,\text{kPa}$, $u(x_{1/2}) = 0.002$, $u(x_{1/3}) = 0.002$, and $u(x_{1/4}) = 0.005$. Solid lines represent phase equilibrium data calculated from PCP-SAFT based on the corresponding pure component and binary interaction parameters shown in Table 5.2 and Table 5.3. For comparison, we show the phase behavior predicted by PCP-SAFT based on pure component parameters only ($k_{ij} = 0$, dashed lines).

rameters are fitted based on one liquid density (Table 5.1) and all pure component vapor pressure data (Table C.2) obtained in this Chapter. The resulting pure component PCP-SAFT parameters are listed in Table 5.2. Binary interaction parameters for the dispersion interaction $k_{ij,\text{binary}}$ are fitted to the respective binary pTx-data sets (Table C.3). The resulting binary interaction parameters $k_{ij,\text{binary}}$ are listed in Table 5.3.

As shown in Figures 5.2, 5.3 and 5.4, the model agrees well with our experimental data. The quality of fit is assessed based on the relative deviations in terms of pressure Δp_{fit} (see Equation 5.1) of experimentally determined data from the calculated phase equilibrium data. For each binary system, we determine the deviations in terms of pressure for a given temperature and liquid phase composition:

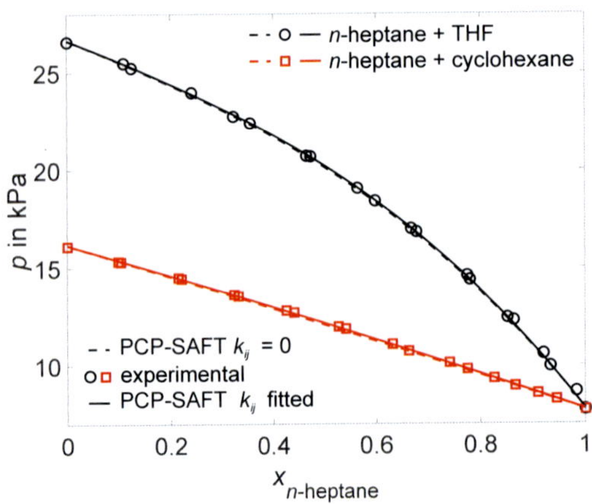

Figure 5.3: Isothermal pressure - liquid-phase composition ($p - x$)-diagram representing the vapor-liquid equilibrium of n-heptane (2) + THF (3) and n-heptane + cyclohexane (4) at $T = 303.2\,\mathrm{K}$. The measurement uncertainties are $u(T) = 0.2\,\mathrm{K}$, $u(p) = 0.2\,\mathrm{kPa}$, $u(x_{2/3}) = 0.002$, and $u(x_{2/4}) = 0.006$. Solid lines represent phase equilibrium data calculated from PCP-SAFT based on the corresponding pure component and binary interaction parameters shown in Table 5.2 and Table 5.3. For comparison, we show the phase behavior predicted by PCP-SAFT based on pure component parameters only ($k_{ij} = 0$, dashed lines).

$$\Delta p = \frac{1}{N}\sum_{i=1}^{N} \Delta p_i = \frac{1}{N}\sum_{i=1}^{N}\left|\frac{p_i^{\mathrm{exp}} - p_i^{\mathrm{cal}}}{p_i^{\mathrm{exp}}}\right| \tag{5.1}$$

with N, the number of experimentally determined pressures, the superscript exp indicates experimental data, and the superscript cal indicates data calculated from PCP-SAFT.

We find deviations ranging from $\Delta p_{\mathrm{fit}} = 0.2\,\%$ to up to $\Delta p_{\mathrm{fit}} = 0.8\,\%$, indicating a good quality of fit for all binary systems studied in this work (see Table 5.3).

For comparison, we predict the binary phase behavior based on pure component data only: for this purpose, we set the binary interaction coefficients k_{ij} to zero. The resulting predicted mixture behavior (dashed lines) is shown in comparison to the fitted mixture behavior (solid lines) for the binary phase equilibria in Figures 5.2, 5.3 and 5.4. The quality of the prediction is discussed in the following paragraphs.

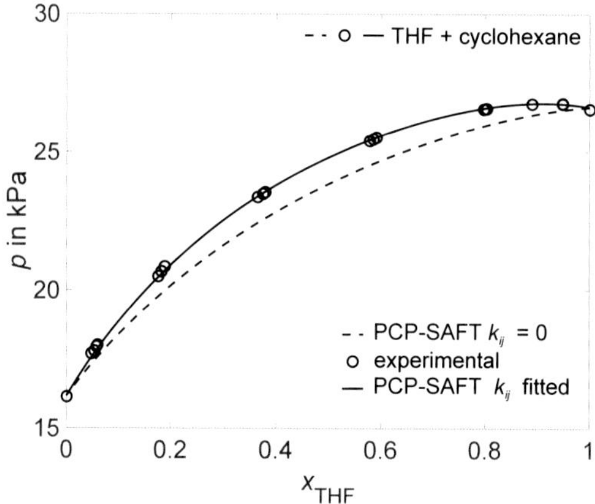

Figure 5.4: Isothermal pressure - liquid-phase composition ($p-x$)-diagram representing the vapor-liquid equilibrium of THF (3) + cyclohexane (4) at T = 303.2 K. The measurement uncertainties are $u(T) = 0.2\,\mathrm{K}$, $u(p) = 0.2\,\mathrm{kPa}$, and $u(x_{3/4}) = 0.002$. Solid lines represent phase equilibrium data calculated from PCP-SAFT based on the corresponding pure component and binary interaction parameters shown in Table 5.2 and Table 5.3. For comparison, we show the phase behavior predicted by PCP-SAFT based on pure component parameters only ($k_{ij} = 0$, dashed lines).

The binary systems of 2-butanone + THF, n-heptane + THF, n-heptane + cyclohexane deviate only slightly from ideal mixture behavior (see Figure 5.2 and Figure 5.3). For these three close to ideal binary systems, we find a good qualitative agreement of experimental data and the prediction from pure component parameters. To evaluate the agreement quantitatively, we calculate the deviation of experimental and predicted data in terms of pressure (Equation 5.1) with a binary interaction parameter k_{ij} set to zero ($\Delta p_{k_{ij}=0}$, see Table 5.3). We find deviations ranging from $\Delta p_{k_{ij}=0}$ = 0.5 % to $\Delta p_{k_{ij}=0}$ = 0.8 %, indicating excellent prediction from pure component parameters only.

We find more pronounced deviations from ideal mixture behavior, and even azeotropic behavior, for the binary systems of 2-butanone + n-heptane, 2-butanone + cyclohexane and THF + cyclohexane (see Figure 5.2 and Figure 5.4). For the binary systems of 2-butanone + n-heptane and 2-butanone + cyclohexane, we still find a good qualitative agreement of the prediction from pure components and the experimental data,

Table 5.3: Binary interaction parameters $k_{ij,\text{binary}}$ fitted to binary experimental data and resulting deviation in terms of pressure (Δp_{fit}) of experimentally determined data from the calculated phase equilibrium data for all binary mixtures studied in this work. For comparison, the deviation in terms of pressure is also calculated with a binary interaction parameter k_{ij} set to zero ($\Delta p_{k_{ij}=0}$).

Mixture	Binary interaction parameter $k_{ij,\text{binary}}$	Δp_{fit} / %	$\Delta p_{k_{ij}=0}$ / %
2-butanone + n-heptane	$5.745 \cdot 10^{-3}$	0.8	3.6
2-butanone + THF	$-1.920 \cdot 10^{-3}$	0.2	0.8
2-butanone + cyclohexane	$1.238 \cdot 10^{-2}$	0.8	6.7
n-heptane + THF	$6.021 \cdot 10^{-4}$	0.6	0.7
n-heptane + cyclohexane	$9.961 \cdot 10^{-4}$	0.2	0.5
THF + cyclohexane	$6.291 \cdot 10^{-3}$	0.2	2.7

as the azeotropic behavior is predicted (Figure 5.2). For the THF + cyclohexane system, however, the qualitative agreement is poor: no azeotrope is predicted based on pure component parameters only (see Figure 5.4). Again, to evaluate the agreement of experimental data and prediction, we calculate the deviation in terms of pressure (Equation 5.1) with a binary interaction parameter k_{ij} set to zero ($\Delta p_{k_{ij}=0}$, see Table 5.3). We find deviations ranging from $\Delta p_{k_{ij}=0} = 2.7\,\%$ to $\Delta p_{k_{ij}=0} = 6.7\,\%$, showing that the prediction of more complex, i.e. azeotropic, phase behavior based on pure component parameters only, is more challenging than for close to ideal systems.

Additionally, we use our phase equilibrium calculations based on PCP-SAFT to compare our data to literature data. Our experimental data and the experimental data from literature are not directly comparable as the literature data were not collected at the same conditions that are chosen in this work (isothermal, $T = 303.2\,\text{K}$). To still be able to evaluate the agreement of our data and data from literature, we use our PCP-SAFT parameters for extrapolation to literature conditions. We calculate the absolute average deviation in terms of pressure $\Delta p_{\text{literature}}$ (see Equation 5.1) of literature data and data calculated based on our PCP-SAFT parameters. Again, pressures are calculated for a given temperature and liquid phase composition. The resulting deviations are listed in Table 5.4. For each binary system, we find at least one data set from literature that shows very good agreement ($\Delta p_{\text{literature}} < 1\,\%$) with the calculated phase equilibrium data. The good agreement is particularly remarkable as conditions from literature (e.g. isobaric, 94 kPa) differ significantly from conditions in this work (isothermal, 303.2 K). Based on the comparison to literature data, we find no clear trend that might indicate systematical errors in our data. The low deviations indicate both the very good extrapolation capabilities of PCP-SAFT and the high quality of our experimental data.

Table 5.4: Absolute average deviations in terms of pressure ($\Delta p_{\text{literature}}$) of experimental data from literature and data calculated from PCP-SAFT based on the corresponding parameters shown in Table 5.2 and Table 5.3. Experimental data from literature are not directly comparable to our experimental data as they were not collected at the conditions studied in this Chapter (isothermal, 303.2 K). The conditions studied in the literature are indicated in the column "Constant value" (isothermal, isobaric, azeotrope) and "Resulting p/T-range".

2-butanone (1) + n-heptane (2)				
Reference	Constant value	Resulting p/T-range	No. of data points	$\Delta p_{\text{literature}}$ / %
(Takeo et al., 1979)	$T = 318.15\,\text{K}$	$p = 15.344 - 31.919\,\text{kPa}$	18	0.63
(Wisniak et al., 1998)	$p = 94\,\text{kPa}$	$T = 348.13 - 369.02\,\text{K}$	27	1.64
(Lladosa et al., 2011)	$p = 20\,\text{kPa}$	$T = 306.59 - 324.56\,\text{K}$	16	2.3
(Lladosa et al., 2011)	$p = 101.3\,\text{kPa}$	$T = 350.35 - 371.35\,\text{K}$	15	1.54
(Steinhauser and White, 1949)	$p = 101.325\,\text{kPa}$	$T = 350.15 - 371.45\,\text{K}$	19	2.07

2-butanone (1) + THF(3)				
Reference	Constant value	Resulting p/T-range	No. of data points	$\Delta p_{\text{literature}}$ / %
(Wisniak et al., 1998)	$p = 94\,\text{kPa}$	$T = 336.82 - 350.42\,\text{K}$	21	0.79
(Vittal Prasad et al., 2007)	$p = 94.7\,\text{kPa}$	$T = 337.55 - 351.25\,\text{K}$	8	1.77

(continued on next page)

Table 5.4 (continued)

2-butanone (1) + cyclohexane (4)				
Reference	Constant value	Resulting p/T-range	No. of data points	$\Delta p_{\text{literature}}$ / %
(Yan et al., 2001)	azeotrope	$p = 101.3 - 902\,\text{kPa}$ $T = 344.88 - 434.57\,\text{K}$	5	0.83
(Crespo Colin et al., 1984)	$T = 323.15\,\text{K}$	$p = 36.066 - 49.257\,\text{kPa}$	32	0.37
(van Nhu and Kohler, 1989)	$T = 293.15\,\text{K}$	$p = 9.481 - 14.404\,\text{kPa}$	21	0.72
(van Nhu and Kohler, 1989)	$T = 313.15\,\text{K}$	$p = 23.659 - 33.539\,\text{kPa}$	28	0.50

n-heptane (2) + THF (3)				
Reference	Constant value	Resulting p/T-range	No. of data points	$\Delta p_{\text{literature}}$ / %
(Deshpande and Oswal, 1975)	$T = 298.15\,\text{K}$	$p = 6.01 - 21.09\,\text{kPa}$	15	0.52
(Deshpande and Oswal, 1975)	$T = 313.15\,\text{K}$	$p = 12.32 - 39.58\,\text{kPa}$	15	0.61
(Wisniak et al., 1997)	$p = 94\,\text{kPa}$	$T = 336.70 - 368.55\,\text{K}$	21	1.14
(Du et al., 2001)	$p = 101.3\,\text{kPa}$	$T = 339.12 - 371.51\,\text{K}$	26	1.11

(continued on next page)

Table 5.4 (continued)

n-heptane (2) + cyclohexane (4)				
Reference	Constant value	Resulting p/T-range	No. of data points	$\Delta p_{\text{literature}}$ / %
(Katayama et al., 1965)	$T = 298.15\,\text{K}$	$p = 6.397 - 12.669\,\text{kPa}$	11	1.06
(Chamorro et al., 2004)	$T = 313.15\,\text{K}$	$p = 12.329 - 24.635\,\text{kPa}$	25	0.15
(Jan et al., 1994)	$p = 101.1\,\text{kPa}$	$T = 353.79 - 371.45\,\text{K}$	16	0.64
(Martin and Youings, 1980)	$T = 298.15\,\text{K}$	$p = 6.095 - 13.004\,\text{kPa}$	32	0.68
(Rodrigues et al., 2005)	$p = 100.65\,\text{kPa}$	$T = 353.78 - 371.42\,\text{K}$	17	1.18
(Young et al., 1977)	$T = 298.15\,\text{K}$	$p = 6.78 - 12.26\,\text{kPa}$	11	1.69
(Sieg, 1950)	$p = 101.325\,\text{kPa}$	$T = 353.90 - 371.58\,\text{K}$	19	1.37
(Lozano et al., 1995)	$T = 298.15\,\text{K}$	$p = 6.114 - 13.007\,\text{kPa}$	21	0.81
(Lozano et al., 1995)	$T = 313.15\,\text{K}$	$p = 12.344 - 24.631\,\text{kPa}$	22	0.29
(Saez et al., 1985)	$T = 333.15\,\text{K}$	$p = 30.823 - 51.517\,\text{kPa}$	17	0.43

THF (3) + cyclohexane (4)				
Reference	Constant value	Resulting p/T-range	No. of data points	$\Delta p_{\text{literature}}$ / %
(Zhang et al., 2013)	$p = 101.3\,\text{kPa}$	$T = 339.04 - 353.83\,\text{K}$	22	0.83
(Deshpande and Oswal, 1975)	$T = 298.15\,\text{K}$	$p = 13.01 - 21.76\,\text{kPa}$	16	0.94
(Deshpande and Oswal, 1975)	$T = 313.15\,\text{K}$	$p = 24.69 - 40.60\,\text{kPa}$	16	1.37

5.5 Quaternary System

Isothermal quaternary vapor-liquid equilibria are characterized using the RAMSPEQU setup. The experimentally characterized VLE data of the quaternary mixture of 2-butanone + n-heptane + THF + cyclohexane at T = 303.2 K are listed in Table C.4. The presented set of quaternary phase equilibrium data (69 pTx-data sets) is measured using a total amount of less than 105 ml of the quaternary mixture (including calibration). Again, the reported uncertainty in the composition $u(x_i)$ (see Table C.4) is the RMSECV from calibration where we use leave-one-out cross validation (see 4.5.3). The values for the uncertainty in composition are in the range of $0.003 < u(x_i) < 0.005$. Thus, the uncertainty in composition for the quaternary system does not differ significantly from the values we found for the binary systems ($0.002 < u(x_{i/j}) < 0.006$, see Section 5.4 and Table C.3). Again, we could not detect a dependence of $u(x_i)$ on composition.

For the model-based analysis of our quaternary measurements, we employ three different sets of binary interaction parameters k_{ij} for PCP-SAFT:

- $k_{ij,\text{quaternary}}$ from a quaternary fit: a new set of binary interaction parameters $k_{ij,\text{quaternary}}$ is obtained by simultaneously fitting to both the binary and quaternary pTx-data sets (see Tables C.3 and C.4),
- $k_{ij,\text{binary}}$ from the binary fit: the binary interaction parameters fitted to binary VLE data (Table 5.3) are used to predict the quaternary VLE.
- $k_{ij} = 0$: the binary interaction parameters are set to zero such that the quaternary VLE is predicted from pure component data only.

In all cases, the accuracy is assessed based on the relative deviations in terms of pressure Δp (Equation 5.1) of experimentally determined quaternary data (p, Table C.4) from the calculated phase equilibrium data ($p_\text{calculation}$, $p_\text{calculation}$, Table C.4) for a given temperature and liquid phase composition.

For the new set of binary interaction parameters $k_{ij,\text{quaternary}}$, the quality of fit is again excellent. We find a deviation of $\Delta p_\text{fit,quaternary} = 0.7\,\%$. The new values for the binary interaction parameters $k_{ij,\text{quaternary}}$ from binary and quaternary data are in good agreement with the $k_{ij,\text{binary}}$ from binary data only (maximum difference of less than 0.001). If the new set of binary interaction parameters $k_{ij,\text{quaternary}}$ are used for the binary phase equilibrium data, we find a maximum difference of 0.2 % from the original deviations in terms of pressure listed in Table 5.3.

For comparison, we also employ the two other sets of binary interaction parameters $k_{ij,\text{binary}}$ and $k_{ij} = 0$ to predict quaternary phase equilibrium data using PCP-SAFT.

For the binary interaction parameters $k_{ij,\text{binary}}$, a deviation of $\Delta p_{\text{prediction}} = 0.9\,\%$ is found. Thus, the deviation between experimentally determined and predicted phase equilibrium data $\Delta p_{\text{prediction}}$ is only slightly larger than the value found for the quality of fit $\Delta p_{\text{fit,quaternary}} = 0.7\,\%$ where the quaternary data had been included in the fit. Hence, the quaternary phase equilibrium is predicted with excellent accuracy using PCP-SAFT parameters adjusted to pure component and binary mixture data only. The good agreement of experimental data and prediction indicates the high quality of both, the prediction using PCP-SAFT and the experimental data.

For further comparison, we employ the set of the binary interaction parameters with k_{ij} set to zero. Thus, predicting the quaternary phase equilibrium data based on pure component data only. We find an average deviation in pressure of $\Delta p_{\text{quaternary},k_{ij}=0}$ $= 3.4\,\%$, which is still good for a prediction of quaternary mixture behavior but significantly worse than the prediction considering binary data.

In summary, quaternary phase equilibrium calculations and experimental data agree best, if all experimental information is taken into account. However, we find that the prediction that uses information on the non-ideality of the binary subsystems, i.e., k_{ij}s fitted to experimental binary data, already yields excellent agreement of phase equilibrium calculations and experimental quaternary data.

5.6 Conclusion

The suitability of the milliliter-scale RAMSPEQU setup is demonstrated for the characterization of the VLE of a quaternary model biofuel. New isothermal VLE data of the quaternary system comprised of 2-butanone + n-heptane + THF + cyclohexane and its binary subsystems at $T = 303.2\,\text{K}$ are presented. To further reduce the experimental effort, we collect pTx-data only.

PCP-SAFT is used for thermodynamic modeling. We find a good quality of fit for the description of the experimental data of the binary subsystems. For comparison, the binary phase behavior is predicted based on pure component parameters only. PCP-SAFT gives very good predictions, even quantitatively, for binary systems that show close to ideal mixture behavior. Significant deviations of predictions and experimental data are found for systems that show more pronounced deviations from ideal mixture behavior, in particular, azeotropic behavior. However, the azeotropic behavior is correctly predicted in two out of three cases. Our PCP-SAFT parameters are used to compare our binary experimental data to binary experimental data from literature at different conditions. Good agreement with data from literature is found, indicating reliable phase equilibrium measurements.

The quaternary phase behavior is measured and compared to data calculated from PCP-SAFT. Similar to the binary subsystems, we find a good quality of fit ($\Delta p_{\text{fit,quaternary}} = 0.7\,\%$). In addition, we predict the phase behavior using parameters adjusted to pure component and binary mixture data only. Quaternary experimental data and prediction agree very well ($\Delta p_{\text{prediction}} = 0.9\,\%$), indicating both the high quality of the prediction using PCP-SAFT and the experimental data. For comparison, the quaternary phase behavior is predicted based on pure component parameters only. The prediction based on pure component parameters is significantly worse ($\Delta p_{\text{quaternary},k_{ij}=0} = 3.4\,\%$), leading to the conclusion that fitting binary interaction parameters is valuable for the prediction of the quaternary phase equilibrium.

The presented binary phase equilibrium data are measured using an average of 22 ml per mixture and the quaternary phase equilibrium data using less than 105 ml of the mixture. In summary, VLE of multicomponent mixtures can reliably and efficiently be characterized using the RAMSPEQU setup.

Chapter 6

Summary and Outlook

6.1 Summary

In this work, we address the need for experimental vapor-liquid equilibrium (VLE) data. Previous VLE characterization methods usually rely on large amounts of substance and require considerable experimental effort (see Section 2.2).

In Chapter 3, the well-established Cailletet method is used to characterize the VLE of the two promising biofuel blends 2-methylfuran + *iso*-octane and 2-methyltetrahydrofuran + di-*n*-butyl ether at high pressure. The Cailletet method serves as a state of the art reference method in the context of this thesis. The crucial steps of sample preparation and phase equilibrium characterization were learned under the supervision of experienced researchers. The Cailletet method is a very sophisticated procedure that was fine-tuned over many years and can therefore produce remarkably accurate results. However, the Cailletet method requires extensive infrastructure, as liquid nitrogen is used in the degassing procedure and the sample is sealed using mercury. Additionally, the handling is complex and time consuming. Immense safety precautions are necessary due to high pressures and temperatures, and particularly handling of mercury under such conditions.

In Chapter 4, the novel RAMSPEQU setup and workflow is presented. In comparison to the Cailletet method, RAMSPEQU cannot attain the same level of accuracy. However, the RAMSPEQU method and workflow requires neither an extensive infrastructure, nor the complex handling and safety precautions. RAMSPEQU overcomes the limitations of previous VLE measurements (see Section 2.2). Phase compositions in both phases are determined by Raman spectroscopic measurements to gain independent composition information while avoiding sampling errors. For validation, vapor pressures of methyl tert-butyl ether (MTBE), ethanol, *iso*-octane and toluene, and the binary VLE of MTBE + *iso*-octane were measured and compared to reference data. Experimental vapor pressures agree with literature data within the measurement uncertainties. We quantified the deviation of our binary VLE data and data from literature using PCP-SAFT for thermodynamic modeling. As we found very good agreement with data from literature ($\Delta p_{\text{literature}} = 0.8\,\%$, $\Delta y_{\text{literature}} = 0.9\,\%$), the RAMSPEQU method was successfully validated.

In-situ degassing was found to be sufficient for phase equilibrium characterization. Therefore, lengthy degassing procedures are not necessary using RAMSPEQU. In-situ degassing reduces the consumption of chemicals, saves time and enables an easy handling of setup and samples. Neither evacuation of the equilibrium cell prior to measurement, nor handling of degassed substances outside of the equilibrium cell is necessary.

Using non-invasive Raman spectroscopy for measurement of phase compositions enables an equilibrium cell volume of less than 3 ml. Consequently, the necessary amount of substance is reduced. Therefore, RAMSPEQU enables analytical VLE measurements for novel substances and their mixtures that are available in small amounts only. The small cell dimensions enhance heat and mass transfer and thereby reduce equilibration time. Additionally, due to small amounts of substance, (potentially) harmful substances can be handled at reduced risk. We further reduce the necessary amount of time and substance by using an integrated workflow for calibration and analysis of phase equilibria. The integrated workflow enables calibration measurements without the need for additional substance. The indirect calibration procedure in the vapor phase avoids a cumbersome and error-prone preparation of vapor samples. Vapor phase compositions agree with the literature, even at the challenging low pressure conditions.

The same mechanism that is used for in-situ degassing can be employed to change the overall composition in the equilibrium cell by removing gaseous substance. Components that are enriched in the vapor phase are removed preferentially. Thus, several compositions can be measured using a single loading of the equilibrium cell. As the remaining sample is already degassed and used to collect another data point, we save time and substance again. Additionally, analysis takes only several seconds as Raman spectra are collected using a highly sensitive CCD sensor.

The combination of the integrated workflow, the accelerated degassing procedure and reduced time for equilibration and analysis leads to rapid data generation. The RAMSPEQU setup gives access to up to 15 $pTxy$-data sets per workday.

The RAMSPEQU setup was validated by characterization of pure component vapor pressures and binary VLE. Even though such types of systems are studied most frequently, systems with only two components rarely depict industrial applications. Instead, VLE data of multicomponent mixtures is of great interest to the chemical industry. Unfortunately, in previous setups, characterization of VLE with more than two components requires considerable experimental effort. The RAMSPEQU method for the efficient characterization of VLE seems particularly suited for the characterization of VLE with more than two components.

In Chapter 5, we characterized the VLE of a quaternary model biofuel and its six binary subsystems using the RAMSPEQU setup. We collected pTx-data only to further reduce the experimental effort. Isothermal VLE data of the quaternary system 2-butanone + n-heptane + THF + cyclohexane and its binary subsystems at $T =$ 303.2 K are reported.

We used PCP-SAFT for thermodynamic modeling and found a good quality of fit for the binary subsystems. Additionally, we compared our binary data through phase equilibrium calculations to experimental data from literature at different conditions. We found an overall low average deviation compared to data from literature indicating the reliability of our binary VLE measurements.

Similarly, for the quaternary phase equilibrium, a good quality of fit is found ($\Delta p_{\text{fit,quaternary}} = 0.7\,\%$). The deviation found for the prediction of the quaternary VLE based on pure component and binary data $\Delta p_{\text{prediction}} = 0.9\,\%$ is only slightly larger than the value for the quality of fit $\Delta p_{\text{fit,quaternary}}$. The good agreement of experimental data and prediction indicates the high quality of both, the prediction using PCP-SAFT and the experimental data collected using RAMSPEQU. The prediction based on pure component data only is significantly worse ($\Delta p_{\text{quaternary},\, k_{ij}=0} = 3.4\,\%$), leading to the conclusion that fitting binary interaction parameters to binary VLE data to gain additional information on the non-ideality of interactions is valuable for the prediction of the quaternary VLE.

The excellent prediction of the quaternary phase behavior from pure component and binary data might lead to the conclusion that measurements on the quaternary system do not yield additional information. However, such an excellent prediction cannot be expected for every system of interest. RAMSPEQU offers the opportunity to gain additional information twofold: first, the quality of a prediction based on systems with a smaller number of components can be assessed by comparison to experimental data. Second, if predictive phase equilibrium calculations for a multicomponent system are not satisfying, RAMSPEQU enables the efficient experimental characterization of the multicomponent VLE.

The binary subsystems were characterized based on an average of 22 ml per mixture, the quaternary data were measured using less than 105 ml of the mixture. Thus, multicomponent VLE were efficiently characterized using the RAMSPEQU setup.

6.2 Perspective for Future Work

In this work, we show that the RAMSPEQU setup can be used to efficiently generate multicomponent VLE data. To further enhance the robustness and efficiency, and extend the applicability of RAMSPEQU to a wider range of phase equilibrium measurements, we suggest potential improvements and estimate the expected effort for the implementation of improvements and the certainty of success.

Fiber optics The measurement of composition in RAMSPEQU employs Raman spectroscopy. For the Raman spectroscopic measurements, laser light is guided into the equilibrium cell from below using several mirrors. The Raman signal is collected via a lens system and guided into a spectrograph for detection. The spectroscopic setup is very sensitive, e.g., to changes in temperature and vibrations. Therefore, laser light irradiation and Raman signal collection constantly need careful maintenance and adjustment. Measurements can only be performed in an optical laboratory. Furthermore, as the laser light from below the equilibrium cell first passes through the liquid phase, the laser light irradiation in the vapor phase changes due to the position of the phase boundary relative to the magnetic stirrer. The changing irradiation can even cause interfering signals from the surrounding water bath and the liquid phase. Other work reports similar challenges (Luther et al., 2015). Due to the necessary maintenance and the changing irradiation, RAMSPEQU needs to be operated by experienced staff. For application in industry, easier handling without the need for constant maintenance and adjustment work is desired. Additionally, influences on irradiation due to passing of the changing phase boundary should be avoided. Therefore, a modification of the setup is suggested: a robust and maintenance-free laser light irradiation and Raman signal collection could be achieved by employing fiber optics. Using fiber optics, single phases could easily be observed separately. For this purpose, fiber optical probes could be used for laser light irradiation and signal collection independently for each phase from the side instead of the bottom of the equilibrium cell. Thereby, light would no longer have to pass through fluid phase boundaries, consequently eliminating the influence of the fluid phase boundary. The setup would not necessarily have to be operated in an optical laboratory. Excluding unforeseeable challenges, we estimate the application of fiber optics in RAMSPEQU to have low risk at high effort. If fiber optics are used for irradiation and signal collection, weaker Raman signals have to be expected in comparison to the current setup. The assessment of low risk at high effort is based on the weaker signal intensity and novel sources of interfering signals that have to be expected and, in turn, eliminated. The anticipated result is a more robust, maintenance-free phase equilibrium characterization method.

Automation The measurement of phase equilibria commonly requires operation of experimental equipment by experienced staff (Hendriks et al., 2010). However, based on the condition that a robust and maintenance-free utilization of RAMSPEQU can be realized, automation of the method might render the need for experienced staff unnecessary. Pending the fiber optical measurement of composition, automation should be straightforward. One necessary instrumental alteration is an electronically operated valve for the degassing respectively changing of composition. Apart from this alteration, automation exclusively requires software to be programmed. Automation could, e.g., be realized by programming a LabView-tool to run the experiment. Currently, successful degassing and equilibration are assessed based on data that are collected using LabView: temperature, pressure and time. Therefore, the criteria for degassing and equilibration can straightforwardly be translated into programming code. The collection of spectra can also be automated as the spectroscopy software offers a LabView interface. Hence, we consider the automation of the RAMSPEQU measurements to exhibit low risk at medium effort. As human interaction is reduced, the anticipated result is an even more efficient phase equilibrium measurement that could also be run overnight. High reproducibility of experimental results is expected.

Vapor-liquid-liquid equilibria In the characterization of vapor-liquid-liquid equilibria (VLLE), methods based on sampling suffer from the same sampling issues as VLE measurements. In dynamic VLLE stills, similar to VLE stills, there are gradients in temperature. Thus, the homogenous vapor phase can split into two liquid phases upon condensation prior to sampling. Consequently, taking representative samples from a VLLE still is challenging. Even if a homogeneous sample can be taken from a system, the sample can still split into two liquid phases prior to analysis, resulting in errors in the analytical chain. These challenges in the measurement of phase compositions can be avoided by directly determining phase compositions in equilibrium using a non-invasive measurement method such as RAMSPEQU.

In addition to the aggravated sampling issues, VLLE equilibration cannot be guaranteed by constant pressure and temperature only. Instead, constant compositions in the respective phases have to be measured. However, sampling commonly changes the overall composition and as a result the equilibrium. Therefore, the non-invasive RAMSPEQU method should be particularly advantageous for the analysis of VLLE as it enables the observation of compositions over time without influencing the equilibrium. Automated measurement of phase equilibria would be advantageous for the time-resolved measurements. Furthermore, in comparison to VLE, an additional phase boundary is present in VLLE. Consequently, employing fiber optics to avoid any influence of the phase boundaries on composition measurements should be particularly beneficial for VLLE.

In particular if VLLE are considered, aqueous systems are of interest. The Raman spectroscopic determination of water mole fractions in a mixture with organic compounds is challenging as water has a comparatively small Raman cross section. However, there are approaches to determine water concentration in organic systems even at high dilution, e.g., by Giraudet et al. (2017); Beumers (2018). Still, the measurement of water contents in organic systems is not an easy task. Therefore, we assess the extension of RAMSPEQU to VLLE to exhibit high risk and high effort. However, if successful, RAMSPEQU should be of very high value for the characterization of VLLE, even for aqueous systems.

In summary, the milliliter-scale **Ram**an **S**pectroscopic **P**hase **Equ**ilibrium Characterization (RAMSPEQU) setup for characterization of VLE was developed and validated. The RAMSPEQU setup and its integrated workflow enable the characterization of multicomponent VLE while achieving significant savings in laboratory time and consumption of chemicals.

Appendix A

Binary high pressure VLE using a Cailletet setup

A.1 Experimental

Table A.1: List of chemicals used in Chapter 3. All chemicals were used as received and no additional purification steps were performed.

substance	provider	purity / wt%	stabilizer (BHT) / ppm
2-MF	Sigma-Aldrich	$\geq$ 98.5	240
2-MTHF	Alfa Aesar	$\geq$ 99.0	150 - 400
DNBE	Sigma-Aldrich	$\geq$ 99.3	-
iso-octane	Merck KGaA	$\geq$ 99.8	-

Table A.2: Bubble-point pressure (p) as function of temperature (T) and mole fraction (x) for the mixture [x 2-MF + (1 - x) *iso*-octane] as measured by the Cailletet technique with the standard uncertainties $u(i)$.

x	T in K	p in Mpa	x	T in K	p in Mpa	x	T in K	p in Mpa
0.000	388.02	0.163	0.100	433.12	0.526	0.200	473.24	1.194
0.000	392.98	0.184	0.100	438.15	0.578	0.200	483.22	1.390
0.000	398.09	0.208	0.100	443.12	0.633	0.200	493.25	1.609
0.000	403.09	0.233	0.100	448.15	0.692	0.200	503.23	1.853
0.000	408.08	0.262	0.100	453.18	0.755	0.200	508.21	1.985
0.000	413.12	0.293	0.100	458.13	0.823	0.300	367.99	0.161
0.000	418.11	0.326	0.100	463.17	0.895	0.300	372.95	0.183
0.000	423.15	0.362	0.100	468.14	0.971	0.300	377.97	0.208
0.000	428.13	0.401	0.100	473.13	1.053	0.300	383.00	0.235
0.000	433.09	0.442	0.100	483.10	1.230	0.300	388.01	0.265
0.000	438.15	0.488	0.100	493.10	1.430	0.300	393.05	0.297
0.000	443.10	0.536	0.100	503.10	1.653	0.300	398.06	0.332
0.000	448.13	0.589	0.100	508.10	1.775	0.300	403.14	0.371
0.000	453.09	0.644	0.200	373.02	0.170	0.300	408.13	0.412
0.000	458.11	0.704	0.200	378.08	0.192	0.300	413.15	0.456
0.000	463.05	0.769	0.200	383.08	0.216	0.300	418.15	0.504
0.000	468.08	0.837	0.200	388.11	0.242	0.300	423.12	0.556
0.000	473.08	0.911	0.200	393.11	0.271	0.300	428.13	0.611
0.000	483.10	1.072	0.200	398.21	0.303	0.300	433.13	0.671
0.000	493.14	1.255	0.200	403.17	0.337	0.300	438.13	0.736
0.000	503.18	1.459	0.200	408.13	0.374	0.300	443.13	0.804
0.000	508.17	1.571	0.200	413.15	0.413	0.300	448.11	0.877
0.100	377.93	0.160	0.200	418.16	0.457	0.300	453.08	0.954
0.100	382.95	0.180	0.200	423.15	0.503	0.300	458.11	1.037
0.100	387.96	0.203	0.200	428.14	0.553	0.300	463.10	1.124
0.100	393.02	0.228	0.200	433.15	0.607	0.300	468.12	1.218
0.100	398.07	0.256	0.200	438.18	0.665	0.300	473.10	1.317
0.100	403.11	0.286	0.200	443.19	0.728	0.300	483.11	1.532
0.100	408.06	0.318	0.200	448.17	0.793	0.300	493.08	1.771
0.100	413.13	0.353	0.200	453.22	0.864	0.300	503.06	2.038
0.100	418.11	0.391	0.200	458.21	0.939	0.300	508.06	2.182
0.100	423.11	0.432	0.200	463.22	1.019	0.404	362.98	0.158
0.100	428.15	0.477	0.200	468.22	1.104	0.404	368.06	0.180

(continued on next page)

Table A.2 (continued)

x	T in K	p in Mpa	x	T in K	p in Mpa	x	T in K	p in Mpa
0.404	373.09	0.204	0.500	398.14	0.407	0.600	423.20	0.724
0.404	377.97	0.230	0.500	403.19	0.455	0.600	428.25	0.798
0.404	383.05	0.260	0.500	408.21	0.505	0.600	433.22	0.875
0.404	388.06	0.293	0.500	413.21	0.559	0.600	438.23	0.958
0.404	392.98	0.329	0.500	418.21	0.617	0.600	443.23	1.046
0.404	398.09	0.367	0.500	423.17	0.679	0.600	448.27	1.141
0.404	403.17	0.411	0.500	428.20	0.747	0.600	453.25	1.241
0.404	408.18	0.456	0.500	433.19	0.818	0.600	458.22	1.347
0.404	413.18	0.506	0.500	438.21	0.895	0.600	463.23	1.460
0.404	418.17	0.558	0.500	443.10	0.976	0.600	468.27	1.582
0.404	423.16	0.615	0.500	448.16	1.064	0.600	473.28	1.709
0.404	428.17	0.677	0.500	453.15	1.157	0.600	483.26	1.984
0.404	433.17	0.742	0.500	458.16	1.256	0.600	493.29	2.293
0.404	438.18	0.813	0.500	463.15	1.362	0.600	503.33	2.635
0.404	443.16	0.888	0.500	468.17	1.473	0.600	508.32	2.820
0.404	448.22	0.969	0.500	473.17	1.591	0.700	357.85	0.169
0.404	453.22	1.055	0.500	483.18	1.846	0.700	362.84	0.194
0.404	458.21	1.146	0.500	493.18	2.132	0.700	367.82	0.223
0.404	463.22	1.242	0.500	503.13	2.447	0.700	372.83	0.254
0.404	468.20	1.343	0.500	508.17	2.618	0.700	377.80	0.287
0.404	473.18	1.451	0.600	357.99	0.159	0.700	382.81	0.325
0.404	483.19	1.688	0.600	363.02	0.183	0.700	387.83	0.366
0.404	493.22	1.950	0.600	367.99	0.209	0.700	392.85	0.411
0.404	503.20	2.240	0.600	372.98	0.238	0.700	397.96	0.460
0.404	508.21	2.397	0.600	378.07	0.271	0.700	402.98	0.514
0.500	358.12	0.153	0.600	383.09	0.306	0.700	407.99	0.572
0.500	363.12	0.175	0.600	388.13	0.345	0.700	413.00	0.633
0.500	368.09	0.200	0.600	393.11	0.387	0.700	418.00	0.701
0.500	373.07	0.227	0.600	398.18	0.433	0.700	422.99	0.773
0.500	378.10	0.257	0.600	403.24	0.484	0.700	427.90	0.848
0.500	383.10	0.289	0.600	408.25	0.537	0.700	432.90	0.932
0.500	388.10	0.325	0.600	413.22	0.595	0.700	437.94	1.021
0.500	393.11	0.364	0.600	418.27	0.658	0.700	442.95	1.116

(continued on next page)

Table A.2 (continued)

x	T in K	p in Mpa	x	T in K	p in Mpa	x	T in K	p in Mpa
0.700	447.96	1.217	0.800	468.05	1.808	0.900	503.17	3.240
0.700	452.98	1.325	0.800	473.02	1.954	0.900	508.17	3.473
0.700	458.00	1.440	0.800	483.07	2.276	1.000	353.10	0.167
0.700	463.00	1.561	0.800	493.08	2.634	1.000	358.07	0.194
0.700	468.01	1.691	0.800	503.13	3.034	1.000	363.06	0.224
0.700	473.02	1.828	0.800	508.12	3.251	1.000	368.02	0.257
0.700	483.00	2.125	0.900	353.05	0.164	1.000	372.97	0.294
0.700	493.01	2.456	0.900	358.02	0.190	1.000	377.98	0.334
0.700	502.99	2.825	0.900	363.04	0.219	1.000	382.99	0.379
0.700	508.03	3.027	0.900	368.03	0.251	1.000	387.96	0.427
0.800	352.93	0.155	0.900	373.09	0.286	1.000	392.97	0.480
0.800	357.86	0.179	0.900	378.07	0.325	1.000	397.93	0.538
0.800	362.79	0.205	0.900	383.10	0.367	1.000	402.98	0.600
0.800	367.78	0.235	0.900	388.07	0.413	1.000	407.95	0.670
0.800	372.86	0.268	0.900	393.10	0.465	1.000	412.96	0.743
0.800	377.92	0.305	0.900	398.09	0.520	1.000	417.96	0.823
0.800	382.92	0.346	0.900	403.13	0.580	1.000	422.94	0.909
0.800	387.95	0.390	0.900	408.15	0.646	1.000	427.97	1.001
0.800	392.93	0.437	0.900	413.14	0.716	1.000	432.93	1.099
0.800	398.02	0.490	0.900	418.14	0.792	1.000	437.97	1.206
0.800	403.04	0.547	0.900	423.16	0.875	1.000	442.93	1.319
0.800	408.06	0.609	0.900	428.16	0.963	1.000	447.95	1.442
0.800	413.07	0.675	0.900	433.19	1.058	1.000	452.97	1.571
0.800	418.09	0.748	0.900	438.18	1.160	1.000	457.97	1.710
0.800	423.12	0.825	0.900	443.20	1.268	1.000	462.96	1.856
0.800	428.07	0.907	0.900	448.19	1.383	1.000	467.96	2.013
0.800	433.08	0.995	0.900	453.18	1.506	1.000	472.97	2.179
0.800	438.10	1.091	0.900	458.21	1.639	1.000	482.96	2.541
0.800	443.07	1.192	0.900	463.13	1.776	1.000	492.93	2.947
0.800	448.03	1.301	0.900	468.18	1.927	1.000	502.93	3.403
0.800	453.03	1.416	0.900	473.17	2.083	1.000	507.92	3.651
0.800	458.04	1.539	0.900	483.15	2.426			
0.800	463.03	1.669	0.900	493.14	2.811			

$u(x) = 0.001, u(T) = 0.01\,\mathrm{K}, u(p) = 0.0005\,\mathrm{MPa}$

Table A.3: Bubble-point pressure (p) as function of temperature (T) and mole fraction (x) for the mixture [x 2-MTHF + (1 - x) DNBE] as measured by the Cailletet technique with the standard uncertainties $u(i)$.

x	T in K	p in Mpa	x	T in K	p in Mpa	x	T in K	p in Mpa
0.000	430.15	0.159	0.200	428.11	0.246	0.400	393.05	0.154
0.000	435.15	0.178	0.200	433.10	0.275	0.400	398.15	0.174
0.000	440.15	0.200	0.200	438.06	0.304	0.400	403.16	0.197
0.000	445.15	0.224	0.200	443.02	0.337	0.400	408.19	0.222
0.000	450.15	0.251	0.200	448.04	0.372	0.400	413.16	0.248
0.000	453.15	0.267	0.200	453.04	0.410	0.400	418.20	0.277
0.000	458.15	0.297	0.200	457.94	0.449	0.400	423.18	0.308
0.000	463.15	0.329	0.200	462.94	0.493	0.400	428.18	0.343
0.000	468.15	0.363	0.200	467.95	0.540	0.400	433.17	0.380
0.000	473.15	0.401	0.200	473.00	0.590	0.400	438.15	0.419
0.000	483.15	0.486	0.200	483.02	0.701	0.400	443.18	0.462
0.000	493.15	0.584	0.200	493.01	0.828	0.400	448.18	0.508
0.000	503.15	0.696	0.200	503.01	0.970	0.400	453.17	0.557
0.000	508.15	0.758	0.200	508.02	1.048	0.400	458.17	0.610
0.100	418.13	0.159	0.300	403.00	0.167	0.400	463.19	0.666
0.100	423.14	0.179	0.300	408.04	0.188	0.400	468.18	0.727
0.100	428.12	0.201	0.300	413.07	0.211	0.400	473.20	0.790
0.100	433.11	0.224	0.300	418.03	0.236	0.400	483.15	0.931
0.100	438.15	0.250	0.300	423.03	0.263	0.400	493.21	1.089
0.100	443.12	0.277	0.300	428.09	0.293	0.400	503.20	1.266
0.100	448.12	0.308	0.300	433.06	0.325	0.400	508.24	1.363
0.100	453.14	0.340	0.300	438.06	0.360	0.500	392.98	0.177
0.100	458.12	0.375	0.300	443.05	0.397	0.500	398.06	0.200
0.100	463.14	0.413	0.300	448.05	0.438	0.500	403.09	0.225
0.100	468.15	0.454	0.300	453.04	0.481	0.500	408.08	0.254
0.100	473.11	0.498	0.300	458.05	0.527	0.500	413.09	0.284
0.100	483.15	0.597	0.300	463.05	0.577	0.500	418.10	0.316
0.100	493.17	0.709	0.300	468.07	0.630	0.500	423.15	0.352
0.100	503.15	0.836	0.300	473.04	0.687	0.500	428.10	0.390
0.100	508.13	0.906	0.300	483.04	0.811	0.500	433.10	0.431
0.200	413.10	0.176	0.300	493.03	0.952	0.500	438.09	0.475
0.200	418.11	0.197	0.300	503.01	1.111	0.500	443.07	0.522
0.200	423.13	0.221	0.300	508.03	1.198	0.500	448.06	0.574

(continued on next page)

Table A.3 (continued)

x	T in K	p in Mpa	x	T in K	p in Mpa	x	T in K	p in Mpa
0.500	453.05	0.628	0.702	387.97	0.200	0.800	433.13	0.609
0.500	458.03	0.687	0.702	392.99	0.226	0.800	438.13	0.670
0.500	463.01	0.748	0.702	398.02	0.255	0.800	443.11	0.736
0.500	467.97	0.815	0.702	403.05	0.287	0.800	448.10	0.806
0.500	473.01	0.887	0.702	408.06	0.321	0.800	453.11	0.882
0.500	482.96	1.041	0.702	413.06	0.359	0.800	458.13	0.962
0.500	492.96	1.216	0.702	418.05	0.399	0.800	463.16	1.050
0.500	502.94	1.410	0.702	423.08	0.444	0.800	468.16	1.141
0.500	507.98	1.516	0.702	428.08	0.492	0.800	473.14	1.238
0.600	383.14	0.158	0.702	433.07	0.543	0.800	483.16	1.450
0.600	388.13	0.179	0.702	438.00	0.597	0.800	493.13	1.688
0.600	393.15	0.203	0.702	443.04	0.657	0.800	503.16	1.952
0.600	398.23	0.229	0.702	448.05	0.720	0.800	508.14	2.096
0.600	403.25	0.257	0.702	453.07	0.788	0.900	373.10	0.166
0.600	408.25	0.288	0.702	458.06	0.860	0.900	378.09	0.190
0.600	413.26	0.322	0.702	463.06	0.937	0.900	383.12	0.216
0.600	418.27	0.359	0.702	468.08	1.019	0.900	388.13	0.246
0.600	423.27	0.398	0.702	473.04	1.108	0.900	393.13	0.278
0.600	428.27	0.442	0.702	483.05	1.299	0.900	398.19	0.314
0.600	433.28	0.489	0.702	493.05	1.511	0.900	403.23	0.353
0.600	438.29	0.539	0.702	503.04	1.749	0.900	408.25	0.395
0.600	443.25	0.591	0.702	508.06	1.878	0.900	413.24	0.441
0.600	448.27	0.649	0.800	378.02	0.173	0.900	418.27	0.491
0.600	453.27	0.710	0.800	383.01	0.197	0.900	423.26	0.544
0.600	458.27	0.776	0.800	388.04	0.224	0.900	428.24	0.602
0.600	463.29	0.846	0.800	393.00	0.254	0.900	433.25	0.664
0.600	468.30	0.920	0.800	398.01	0.286	0.900	438.25	0.731
0.600	473.27	0.999	0.800	403.12	0.322	0.900	443.25	0.804
0.600	483.28	1.171	0.800	408.15	0.361	0.900	448.24	0.881
0.600	493.30	1.366	0.800	413.11	0.404	0.900	453.26	0.964
0.600	503.28	1.582	0.800	418.09	0.449	0.900	458.28	1.052
0.600	508.27	1.698	0.800	423.15	0.499	0.900	463.28	1.146
0.702	382.97	0.176	0.800	428.12	0.551	0.900	468.28	1.245

(continued on next page)

Table A.3 (continued)

x	T in K	p in Mpa
0.900	473.28	1.351
0.900	483.31	1.583
0.900	493.31	1.844
0.900	503.30	2.133
0.900	508.28	2.290
1.000	368.95	0.163
1.000	372.89	0.181
1.000	377.88	0.207
1.000	382.89	0.237
1.000	387.86	0.268
1.000	392.93	0.304
1.000	398.03	0.344
1.000	403.02	0.386
1.000	408.05	0.432
1.000	413.07	0.483
1.000	418.08	0.538
1.000	423.11	0.597
1.000	428.18	0.661
1.000	433.16	0.730
1.000	438.19	0.804
1.000	443.19	0.883
1.000	448.17	0.968
1.000	453.22	1.060
1.000	458.19	1.157
1.000	463.25	1.262
1.000	468.25	1.372
1.000	473.20	1.489
1.000	483.25	1.746
1.000	493.28	2.035
1.000	503.24	2.358
1.000	508.21	2.532

$u(x) = 0.001$
$u(T) = 0.01\,\mathrm{K}$
$u(p) = 0.0005\,\mathrm{MPa}$

Appendix B

RAMSPEQU: A milliliter-scale setup for VLE

B.1 Experimental

Table B.1: List of chemicals used in Chapter 4. All chemicals were used as received and no additional purification steps were performed.

substance	provider	purity / wt%
MTBE	VWR	≥ 99.8
iso-octane	Merck Millipore	≥ 99.8
ethanol	Merck Millipore	≥ 99.9
toluene	VWR	≥ 99.8
nitrogen	Westfalengas	≥ 99.999

Table B.2: Experimentally determined pure component densities (ρ) and measurement uncertainties $u(\rho)$ at $p = 100\,\text{kPa}$ and $T = 298.15\,\text{K}$ for MTBE (Landaverde-Cortes et al., 2007) and *iso*-octane (Pádua et al., 1996) from literature.

substance	$\rho/\text{g}\,\text{cm}^{-3}$	$u(\rho)/\text{g}\,\text{cm}^{-3}$
MTBE	0.7354	0.00003
iso-octane	0.6878	0.0003

Table B.3: Experimental vapor pressure p at the corresponding temperature T for MTBE, ethanol, *iso*-octane and toluene with the standard uncertainties $u(i)$ as collected with the RAMSPEQU setup.

MTBE		ethanol		*iso*-octane		toluene	
T / K	p / kPa	T / K	p / kPa	T / K	p / kPa	T / K	p / kPa
283.6	17.79	278.6	2.18	283.6	3.05	278.7	1.26
288.6	22.07	283.5	3.11	288.4	3.99	283.6	1.70
293.5	27.43	293.4	5.95	293.4	5.18	288.5	2.26
298.4	33.72	298.3	7.96	298.3	6.63	293.4	2.97
303.3	41.10	303.2	10.52	298.4	6.62	298.3	3.89
308.2	49.80	308.1	13.84	303.3	8.39	303.2	5.00
313.1	59.90	313.1	17.87	308.2	10.53	308.1	6.27
318.1	71.61	318.1	22.98	313.1	12.97	313.1	7.99
323.0	85.01	322.9	29.05	318.1	15.96	318.0	9.95
327.9	100.34	327.9	36.73	323.0	19.54	323.0	12.26
332.9	117.70	332.8	46.15	328.0	23.81	327.9	15.00
337.8	137.39			332.8	28.66	332.8	18.26
						337.8	22.04

$u(T) = 0.2\,\mathrm{K}, u(p) = 0.2\,\mathrm{kPa}$

Table B.4: Isothermal $pTxy$-data at 318.1 K for the mixture MTBE (1) + *iso*-octane (2) with the standard uncertainties $u(i)$ as collected with the RAMSPEQU setup.

p / kPa	x_1	y_1
15.96	0.000	0.00
19.69	0.058	0.24
20.21	0.064	0.26
20.66	0.069	0.27
21.09	0.074	0.28
21.52	0.088	0.31
25.07	0.140	0.41
26.12	0.159	0.45
29.65	0.218	0.58
31.91	0.254	0.62
33.28	0.277	0.65
37.94	0.355	0.71
41.08	0.410	0.74
44.03	0.470	0.81
46.93	0.526	0.83
54.99	0.684	0.90
57.32	0.728	0.92
58.89	0.757	0.93
66.41	0.893	0.96
67.21	0.906	0.99
67.58	0.914	0.99
71.61	1.000	1.00

$u(T) = 0.2\,\text{K}$
$u(p) = 0.2\,\text{kPa}$
$u(x) = 0.008$
$u(y) = 0.01$

Appendix C

VLE of a quaternary model biofuel using RAMSPEQU

C.1 Experimental

Table C.1: List of chemicals used in Chapter 5. All chemicals were used as received and no additional purification steps were performed.

substance	provider	purity / wt%
2-butanone	VWR	≥ 99.5
cyclohexane	VWR	≥ 99.7
n-heptane	Merck Millipore	≥ 99.3
THF	Merck Millipore	≥ 99.9

Table C.2: Experimental vapor pressure $p_{s0,i}$ at the corresponding temperature T for 2-butanone, n-heptane, THF and cyclohexane with the standard uncertainties $u(i)$.

2-butanone		n-heptane		THF		cyclohexane	
p / kPa	T / K	p / kPa	T / K	p / kPa	T / K	p / kPa	T / K
5.8	283.7	2.8	283.7	11.0	283.7	6.5	283.8
9.6	293.5	4.8	293.5	17.4	293.5	10.5	293.6
15.1	303.1	7.7	303.1	26.6	303.1	16.1	303.1
23.7	313.2	12.4	313.2	40.1	313.2	24.6	313.2
35.4	323.0	18.8	323.0	58.0	323.0	36.1	323.0
51.4	332.8	27.9	332.8	81.9	332.8	51.3	332.9

$u(T) = 0.2\,\mathrm{K}, u(p) = 0.2\,\mathrm{kPa}$

Table C.3: Isothermal pTx-data with the standard uncertainties $u(i)$ for all binary mixtures studied in Chapter 5.

2-butanone (1) + n-heptane (2)			2-butanone (1) + THF (3)			2-butanone (1) + cyclohexane (4)		
T/K	p/kPa	x_1	T/K	p/kPa	x_1	T/K	p/kPa	x_1
303.2	7.7	0.000	303.2	26.6	0.000	303.2	16.1	0.000
303.2	11.8	0.091	303.2	25.7	0.104	303.2	19.9	0.082
303.2	12.8	0.133	303.2	25.7	0.106	303.1	20.1	0.094
303.2	13.3	0.157	303.2	25.6	0.107	303.2	20.2	0.098
303.2	13.7	0.178	303.2	25.0	0.182	303.2	20.9	0.141
303.2	14.6	0.239	303.2	25.0	0.184	303.2	21.0	0.152
303.2	15.2	0.301	303.2	24.9	0.189	303.2	21.0	0.152
303.2	15.6	0.346	303.2	24.7	0.197	303.1	21.1	0.160
303.2	16.2	0.482	303.2	23.0	0.370	303.2	21.8	0.259
303.2	16.5	0.542	303.2	22.9	0.379	303.2	21.9	0.281
303.2	16.6	0.574	303.2	22.7	0.394	303.2	21.9	0.287
303.2	16.7	0.684	303.2	20.6	0.589	303.2	22.1	0.395
303.2	16.8	0.691	303.2	20.5	0.593	303.1	22.2	0.403
303.2	16.8	0.695	303.1	20.4	0.603	303.2	22.1	0.530
303.2	16.8	0.803	303.2	20.1	0.623	303.2	21.9	0.616
303.2	16.8	0.803	303.2	18.3	0.779	303.2	21.8	0.624
303.2	16.7	0.811	303.1	18.2	0.787	303.2	21.1	0.740
303.2	16.5	0.895	303.2	17.9	0.801	303.2	20.9	0.753
303.2	16.4	0.897	303.2	15.1	1.000	303.2	20.2	0.816
303.2	16.4	0.906				303.2	19.8	0.837
303.2	15.1	1.000				303.2	18.3	0.912
						303.2	18.1	0.917
						303.2	17.9	0.924
						303.2	15.1	1.000

(continued on next page)

Table C.3 (continued)

n-heptane (2) + THF (3)			*n*-heptane (2) + cyclohexane (4)			THF (3) + cyclohexane (4)		
T/K	p/kPa	x_2	T/K	p/kPa	x_2	T/K	p/kPa	x_3
303.2	26.6	0.000	303.2	16.1	0.000	303.2	16.1	0.000
303.2	25.5	0.109	303.2	15.4	0.098	303.2	17.7	0.048
303.2	25.2	0.124	303.2	15.3	0.103	303.2	17.8	0.054
303.2	24.0	0.241	303.2	14.5	0.215	303.2	17.9	0.059
303.2	22.8	0.322	303.2	14.5	0.221	303.2	18.0	0.061
303.2	22.4	0.355	303.2	13.6	0.323	303.1	20.5	0.176
303.2	20.7	0.464	303.2	13.6	0.331	303.2	20.7	0.183
303.2	20.7	0.472	303.2	12.8	0.425	303.2	20.9	0.188
303.2	19.0	0.564	303.2	12.7	0.440	303.2	23.4	0.365
303.2	18.4	0.598	303.2	12.0	0.527	303.2	23.5	0.374
303.2	17.0	0.668	303.2	11.9	0.541	303.2	23.6	0.378
303.2	16.8	0.678	303.2	11.1	0.631	303.2	25.4	0.578
303.2	14.6	0.775	303.2	10.7	0.663	303.2	25.5	0.585
303.2	14.3	0.780	303.2	10.1	0.741	303.2	25.5	0.591
303.1	12.4	0.852	303.2	9.8	0.775	303.2	26.6	0.797
303.1	12.3	0.864	303.2	9.3	0.826	303.2	26.6	0.802
303.1	10.6	0.921	303.2	8.9	0.867	303.2	26.6	0.804
303.2	9.9	0.934	303.2	8.6	0.910	303.2	26.8	0.889
303.2	8.6	0.983	303.2	8.3	0.946	303.2	26.8	0.890
303.2	7.7	1.000	303.2	7.7	1.000	303.2	26.8	0.949
						303.2	26.8	0.949
						303.2	26.6	1.000

$u(T) = 0.2\,\mathrm{K}$, $u(p) = 0.2\,\mathrm{kPa}$, $u(x_{1/2}) = 0.002$, $u(x_{1/3}) = 0.002$
$u(x_{1/4}) = 0.005$, $u(x_{2/3}) = 0.002$, $u(x_{2/4}) = 0.006$, $u(x_{3/4}) = 0.002$

Table C.4: Isothermal pTx-data for the quaternary mixture 2-butanone (1) + n-heptane (2) + THF (3) + cyclohexane (4) with the standard uncertainties $u(i)$. Additionally, pressure values calculated from PCP-SAFT based on pure component, binary and quaternary data ($p_{\text{calculation}}$) respectively pure component and binary data ($p_{\text{prediction}}$) are presented.

T / K	p / kPa	$p_{\text{calculated}}$ / kPa	$p_{\text{predicted}}$ / kPa	x_1	x_2	x_3
303.2	13.7	13.3	13.3	0.060	0.766	0.074
303.2	15.9	15.6	15.6	0.068	0.668	0.157
303.2	15.5	15.2	15.2	0.074	0.620	0.086
303.2	17.3	17.0	17.1	0.082	0.454	0.099
303.2	17.9	17.9	18.0	0.083	0.552	0.252
303.2	17.8	17.7	17.7	0.083	0.507	0.185
303.2	18.8	18.6	18.6	0.091	0.318	0.113
303.2	15.3	15.0	15.0	0.091	0.690	0.101
303.2	19.4	19.4	19.5	0.095	0.334	0.193
303.2	16.7	16.4	16.4	0.098	0.556	0.106
303.2	18.1	17.8	17.8	0.098	0.414	0.111
303.2	19.9	20.0	20.0	0.100	0.378	0.286
303.2	17.8	17.6	17.7	0.102	0.559	0.214
303.2	20.5	20.4	20.5	0.103	0.143	0.125
303.2	19.6	19.3	19.3	0.107	0.281	0.124
303.2	19.2	19.1	19.1	0.111	0.422	0.225
303.2	19.8	19.7	19.8	0.113	0.437	0.325
303.2	20.6	20.9	20.9	0.115	0.366	0.396
303.2	20.9	20.8	20.8	0.116	0.128	0.130
303.2	21.3	21.6	21.6	0.116	0.154	0.224
303.2	20.4	20.4	20.5	0.118	0.274	0.222
303.2	21.9	21.9	22.0	0.121	0.133	0.241
303.2	22.0	22.3	22.4	0.122	0.166	0.322
303.2	21.2	21.2	21.3	0.122	0.295	0.338
303.2	22.7	23.0	23.0	0.126	0.178	0.449
303.2	21.6	21.7	21.7	0.129	0.303	0.443
303.1	23.8	23.8	23.9	0.132	0.141	0.600
303.2	22.7	22.8	22.9	0.132	0.135	0.347
303.2	23.3	23.5	23.5	0.134	0.139	0.478
303.2	23.4	23.6	23.6	0.137	0.158	0.578
303.2	17.6	17.7	17.7	0.175	0.549	0.169
303.2	17.5	17.4	17.5	0.178	0.512	0.094
303.2	19.1	19.2	19.2	0.192	0.345	0.103
303.2	19.7	19.8	19.8	0.203	0.361	0.202

(continued on next page)

Table C.4 (continued)

T / K	p / kPa	$p_{\text{calculated}}$ / kPa	$p_{\text{predicted}}$ / kPa	x_1	x_2	x_3
303.2	17.2	17.2	17.2	0.208	0.566	0.107
303.2	18.5	18.5	18.6	0.218	0.431	0.114
303.2	21.0	21.1	21.2	0.218	0.160	0.119
303.2	20.0	20.2	20.2	0.219	0.366	0.294
303.2	20.0	20.0	20.0	0.222	0.286	0.118
303.2	21.6	21.7	21.8	0.227	0.171	0.224
303.2	19.1	19.1	19.2	0.229	0.434	0.218
303.2	20.6	20.6	20.6	0.232	0.293	0.232
303.2	21.4	21.4	21.5	0.235	0.136	0.126
303.2	20.8	20.9	20.9	0.240	0.302	0.330
303.2	22.1	22.2	22.3	0.242	0.137	0.241
303.2	22.0	22.2	22.3	0.249	0.172	0.333
303.2	22.6	22.6	22.7	0.253	0.140	0.356
303.2	23.0	23.0	23.1	0.253	0.131	0.492
303.2	22.5	22.7	22.8	0.263	0.150	0.467
303.2	17.3	17.4	17.5	0.272	0.531	0.091
303.2	19.3	19.5	19.5	0.324	0.361	0.200
303.2	19.1	19.3	19.3	0.330	0.348	0.102
303.2	18.3	18.4	18.4	0.334	0.436	0.113
303.2	20.1	20.1	20.2	0.353	0.296	0.228
303.2	19.7	19.8	19.8	0.365	0.290	0.115
303.2	21.1	21.3	21.3	0.372	0.155	0.118
303.2	21.8	21.8	21.9	0.381	0.134	0.244
303.2	21.4	21.5	21.5	0.382	0.134	0.124
303.2	21.4	21.6	21.7	0.388	0.150	0.229
303.2	21.8	21.8	21.9	0.394	0.135	0.347
303.2	21.4	21.6	21.7	0.412	0.143	0.328
303.2	18.8	19.0	19.0	0.480	0.305	0.106
303.2	19.2	19.3	19.3	0.493	0.271	0.118
303.2	20.8	20.9	21.0	0.512	0.133	0.122
303.2	20.9	20.9	21.0	0.517	0.127	0.231
303.2	20.7	20.8	20.9	0.518	0.139	0.117
303.2	20.5	20.7	20.7	0.542	0.128	0.214
303.2	19.9	19.9	20.0	0.639	0.124	0.119
303.2	19.6	19.7	19.8	0.659	0.122	0.110

$u(T) = 0.2\,\text{K}$, $u(p) = 0.2\,\text{kPa}$
$u(x_1) = 0.004$, $u(x_2) = 0.005$, $u(x_3) = 0.003$, $u(x_4) = 0.003$

Bibliography

Abbott, M. M. (1986). Low-pressure phase equilibria: Measurement of VLE. *Fluid Phase Equilibria*, 29:193–207.

Abrams, D. S. and Prausnitz, J. M. (1975). Statistical thermodynamics of liquid mixtures: A new expression for the excess Gibbs energy of partly or completely miscible systems. *AIChE Journal*, 21(1):116–128.

Adami, R., Schuster, J., Liparoti, S., Reverchon, E., Leipertz, A., and Braeuer, A. (2013). A Raman spectroscopic method for the determination of high pressure vapour liquid equilibria. *Fluid Phase Equilibria*, 360:265–273.

Aichlmayr, H. T., Kittelson, D. B., and Zachariah, M. R. (2003). Micro-HCCI combustion: Experimental characterization and development of a detailed chemical kinetic model with coupled piston motion. *Combustion and Flame*, 135(3):227–248.

Albers, K. and Sadowski, G. (2011). Minimal Experimental Data Set Required for Estimating PCP-SAFT Parameters. *Industrial & Engineering Chemistry Research*, 50(20):11746–11754.

Alonso, I., Mozo, I., de la Fuente, Isaías García, González, J. A., and Cobos, J. C. (2010). Thermodynamics of Ketone + Amine Mixtures. Part III. Volumetric and Speed of Sound Data at (293.15, 298.15, and 303.15) K for 2-Butanone + Aniline, + N-Methylaniline, or + Pyridine Systems. *Journal of Chemical and Engineering Data*, 55(12):5400–5405.

Alsmeyer, F., Koß, H.-J., and Marquardt, W. (2004). Indirect Spectral Hard Modeling for the Analysis of Reactive and Interacting Mixtures. *Applied Spectroscopy*, 58(8):975–985.

Alsmeyer, F., Marquardt, W., and Olf, G. (2002). A new method for phase equilibrium measurements in reacting mixtures. *Fluid Phase Equilibria*, 203(1-2):31–51.

Azizian, S. and Bashavard, N. (2008). Surface Tension of Dilute Solutions of Alkanes in Cyclohexanol at Different Temperatures. *Journal of Chemical and Engineering Data*, 53(10):2422–2425.

Battino, R., Banzhof, M., Bogan, M., and Wilhelm, E. (1971). Apparatus for rapid degassing of liquids. Part III. *Analytical Chemistry*, 43(6):806–807.

Bernatová, S. and Wichterle, I. (2001). Isothermal vapour–liquid equilibria in the ternary system tert-butyl methyl ether + tert-butanol + 2,2,4-trimethylpentane and the three binary subsystems. *Fluid Phase Equilibria*, 180(1-2):235–245.

Beumers, P. (2018). *Physically based models for the analysis of Raman spectra*. Dissertation, RWTH Aachen University, Aachen, Germany, Aachen.

Beumers, P., Engel, D., Brands, T., Koß, H.-J., and Bardow, A. (2018). Robust analysis of spectra with strong background signals by First-Derivative Indirect Hard Modeling (FD-IHM). *Chemometrics and Intelligent Laboratory Systems*, 172:1–9.

Bond-Watts, B. B., Bellerose, R. J., and Chang, M. C. Y. (2011). Enzyme mechanism as a kinetic control element for designing synthetic biofuel pathways. *Nature Chemical Biology*, 7(4):222–227.

Borges, G. R., Lucas, M. A., Nunes, R. B. M., Amaral, M. d. J., Franceschi, E., Santos, A. F., and Dariva, C. (2015). Near infrared spectroscopy applied for high-pressure phase behavior measurements. *The Journal of Supercritical Fluids*, 104:221–226.

Boublík, T., Fried, V., Hála, E., and Gaube, J. (1984 // 1985). *The vapour pressures of pure substances*, volume 89. Elsevier Science Pub. Co., Inc., New York, NY.

Carveth, H. R. (1899). The Composition of Mixed Vapors, I. *Journal of Physical Chemistry*, 3(4):193–213.

Chadha, R. and Tripathi, A. D. (1995). Excess Molar Enthalpies of 1,1,2,2-Tetrachloroethane + 2-Methylfuran, + Tetrahydrofuran, + 1,4-Dioxane, and + Cyclopentanone at 308.15 and 318.15 K. *Journal of Chemical and Engineering Data*, 40(3):645–646.

Chamorro, C. R., Martin, M. C., Villamanan, M. A., and Segovia, J. J. (2004). Characterization and modelling of a gasoline containing 1,1-dimethylethyl methyl ether (MTBE), diisopropyl ether (DIPE) or 1,1-dimethylpropyl methyl ether (TAME) as fuel oxygenate based on new isothermal binary vapour-liquid data. *Fluid Phase Equilibria*, 220:103–110.

Chapman, W. G., Gubbins, K. E., Jackson, G., and Radosz, M. (1989). SAFT: Equation-of-State Solution Model for Associating Fluids. *Fluid Phase Equilibria*, 52:31–38.

Chirico, R. D., Frenkel, M., Diky, V. V., Marsh, K. N., and Wilhoit, R. C. (2003). ThermoML-An XML-Based Approach for Storage and Exchange of Experimental and Critically Evaluated Thermophysical and Thermochemical Property Data. 2. Uncertainties. *Journal of Chemical and Engineering Data*, 48(5):1344–1359.

Crespo Colin, A., Compostizo, A., and Díaz Peña, M. (1984). Excess Gibbs energy and excess volume of (cyclohexane + 2-propanone) and of (cyclohexane + 2-butanone). *The Journal of Chemical Thermodynamics*, 16(5):497–502.

Croke, B. F. W. (1995). Removal of Cosmic-Ray Events in Spectroscopic CCD Data. *Publications of the Astronomical Society of the Pacific*, 107(718):1255–1258.

DDBST (2018a). DDBSP: Model Parameter Estimator (accessed 2018 Jun 18) http://www.ddbst.de/model-parameter-estimator.html .

DDBST (2018b). Details About Data Types (accessed 2018 Jun 18) http://www.ddbst.com/ddb-azd.html .

de Bruycker, R., Tran, L.-S., Carstensen, H.-H., Glaude, P.-A., Monge, F., Alzueta, M. U., Battin-Leclerc, F., and van Geem, K. M. (2017). Experimental and modeling study of the pyrolysis and combustion of 2-methyl-tetrahydrofuran. *Combustion and Flame*, 176:409–428.

de Loos, T. W., Van der Kooi, Hedzer J., and Ott, P. L. (1986). Vapor-liquid critical curve of the system ethane + 2-methylpropane. *Journal of Chemical and Engineering Data*, 31(2):166–168.

Deshpande, D. D. and Oswal, S. L. (1975). Thermodynamics of mixtures containing p-dioxan or tetrahydrofuran 1. Excess Gibbs free energies and excess volumes. *The Journal of Chemical Thermodynamics*, 7(2):155–159.

Dillon, H. E. and Penoncello, S. G. (2004). A fundamental equation for calculation of the thermodynamic properties of ethanol. *International Journal of Thermophysics*, 25(2):321–335.

Dohrn, R. (1994). *Berechnung von Phasengleichgewichten.* Grundlagen und Fortschritte der Ingenieurwissenschaften. Vieweg, Braunschweig.

Dohrn, R., Fonseca, J. M. S., and Peper, S. (2012). Experimental Methods for Phase Equilibria at High Pressures. *Annual Review of Chemical and Biomolecular Engineering, Vol 3*, 3:343–367.

Du, T.-B., Tang, M., and Chen, Y.-P. (2001). Vapor–liquid equilibria of the binary mixtures of tetrahydrofuran with 2,2,4-trimethylpentane, methylcyclohexane and n-heptane at 101.3 kPa. *Fluid Phase Equilibria*, 192(1-2):71–83.

Ezeji, T. C., Qureshi, N., and Blaschek, H. P. (2004). Butanol fermentation research: upstream and downstream manipulations. *The Chemical Record*, 4(5):305–314.

Fonseca, J. M., Dohrn, R., and Peper, S. (2011). High-pressure fluid-phase equilibria: Experimental methods and systems investigated (2005 - 2008). *Fluid Phase Equilibria*, 300(1-2):1–69.

Forero G., L. A. and Velásquez J., J. A. (2011). Wagner liquid–vapour pressure equation constants from a simple methodology. *The Journal of Chemical Thermodynamics*, 43(8):1235–1251.

Fredenslund, A., Jones, R. L., and Prausnitz, J. M. (1975). Group-contribution estimation of activity coefficients in nonideal liquid mixtures. *AIChE Journal*, 21(6):1086–1099.

Geilen, F. M. A., Engendahl, B., Harwardt, A., Marquardt, W., Klankermayer, J., and Leitner, W. (2010). Selective and Flexible Transformation of Biomass-Derived Platform Chemicals by a Multifunctional Catalytic System. *Angewandte Chemie International Edition*, 122(32):5642–5646.

Giraudet, C., Papavasileiou, K. D., Rausch, M. H., Chen, J., Kalantar, A., van der Laan, G. P., Economou, I. G., and Fröba, A. P. (2017). Characterization of Water Solubility in n-Octacosane Using Raman Spectroscopy. *The Journal of Physical Chemistry B*, 121(47):10665–10673.

Gmehling, J., Onken, U., and Arlt, W. (1991). *Vapor-liquid equilibrium data collection: Tables and diagrams of data for binary and multicomponent mixtures up to moderate pressures; constants of correlation equations for computer use*, volume 1,1 of *Chemistry data series*. Dechema, Frankfurt am Main, 2. ed., 3. printing edition.

Goral, M., Oracz, P., Skrzecz, A., Bok, A., and Maczyński, A. (2003). Recommended Vapor–Liquid Equilibrium Data. Part 2: Binary Alkanol–Alkane Systems. *Journal of Physical and Chemical Reference Data*, 32(4):1429–1472.

Gross, J. (2005). An equation-of-state contribution for polar components: Quadrupolar molecules. *AIChE Journal*, 51(9):2556–2568.

Gross, J. and Sadowski, G. (2000). Application of perturbation theory to a hard-chain reference fluid: an equation of state for square-well chains. *Fluid Phase Equilibria*, 168(2):183–199.

Gross, J. and Sadowski, G. (2001). Perturbed-Chain SAFT: An Equation of State Based on a Perturbation Theory for Chain Molecules. *Industrial & Engineering Chemistry Research*, 40(4):1244–1260.

Gross, J. and Vrabec, J. (2006). An equation-of-state contribution for polar components: Dipolar molecules. *AIChE Journal*, 52(3):1194–1204.

Hendriks, E., Kontogeorgis, G. M., Dohrn, R., de Hemptinne, J.-C., Economou, I. G., Žilnik, L. F., and Vesovic, V. (2010). Industrial Requirements for Thermodynamics and Transport Properties. *Industrial & Engineering Chemistry Research*, 49(22):11131–11141.

Hoppe, F., Burke, U., Thewes, M., Heufer, A., Kremer, F., and Pischinger, S. (2016). Tailor-Made Fuels from Biomass: Potentials of 2-butanone and 2-methylfuran in direct injection spark ignition engines. *Fuel*, 167:106–117.

Horstmann, S., Gardeler, H., Bölts, R., Zudkevitch, D., and Gmehling, J. (1999). Vapor-Liquid Equilibria and Excess Enthalpy Data for the Binary Systems 2-Methyltetrahydrofuran with 2,2,4-Trimethylpentane (Isooctane), Ethanol, Toluene, Cyclohexane and Methylcyclohexane. *Journal of Chemical and Engineering Data*, 44(5):959–964.

Jan, D.-S., Shiau, H.-Y., and Tsai, F.-N. (1994). Vapor-Liquid Equilibria of n-Hexane + Cyclohexane + n-Heptane and the Three Constituent Binary Systems at 101.0 kPa. *Journal of Chemical and Engineering Data*, 39(3):438–440.

Janssen, A. J., Kremer, F. W., Baron, J. H., Muether, M., Pischinger, S., and Klankermayer, J. (2011). Tailor-Made Fuels from Biomass for Homogeneous Low-Temperature Diesel Combustion. *Energy Fuels*, 25(10):4734–4744.

Jiménez, E., Segade, L., Franjo, C., Casas, H., Legido, J., and Paz Andrade, M. (1998). Viscosity deviations of ternary mixtures di-n-butyl ether+1-propanol+n-octane at several temperatures. *Fluid Phase Equilibria*, 149(1–2):339–358.

Kaiser, T., Voßmerbäumer, C., and Schweiger, G. (1992). A new approach to the determination of fluid phase Equilibria: Concentration measurements by Raman spectroscopy. *Berichte der Bunsengesellschaft für physikalische Chemie*, 96(8):976–980.

Katayama, T., Sung, E. K., and Lightfoot, E. N. (1965). Isothermal activity coefficients for the system cyclohexane - n-heptane - toluene at 25 °C. *AIChE Journal*, 11(5):924–929.

Kay, W. B. and Warzel, F. M. (1951). 2,2,4-Trimethylpentane (Iso-octane). Vapor Pressure, Critical Constants, and Saturated Vapor and Liquid Densities. *Industrial and Engineering Chemistry*, 43(5):1150–1152.

Kim, J. K., Choi, J. H., Park, D. R., and Song, I. K. (2013). Etherification of *n*-Butanol to Di-*n*-Butyl Ether Over Keggin-, Wells-Dawson-, and Preyssler-Type Heteropolyacid Catalysts. *Journal of Nanoscience and Nanotechnology*, 13(12):8121–8126.

Klamt, A. (1995). Conductor-like Screening Model for Real Solvents: A New Approach to the Quantitative Calculation of Solvation Phenomena. *The Journal of Physical Chemistry*, 99:2224–2235.

Klein-Douwel, R., Donkerbroek, A. J., van Vliet, A. P., Boot, M. D., Somers, L., Baert, R., Dam, N. J., and ter Meulen, J. J. (2009). Soot and chemiluminescence in diesel combustion of bio-derived, oxygenated and reference fuels. *Proceedings of the Combustion Institute*, 32(2):2817–2825.

Kobe, K. A., Ravicz, A. E., and Vohra, S. P. (1956). Critical Properties and Vapor Pressures of Some Ethers and Heterocyclic Compounds. *Industrial & Engineering Chemistry Chemical & Engineering Data Series*, 1(1):50–56.

Kontogeorgis, G. M. and Folas, G. K. (2010). *Thermodynamic Models for Industrial Applications: From classical and advanced mixing rules to association theories*, volume 1st ed. John Wiley & Sons, Ltd and Wiley, Chichester, UK.

Landaverde-Cortes, D. C., Estrada-Baltazar, A., Iglesias-Silva, G. A., and Hall, K. R. (2007). Densities and Viscosities of MTBE+ Heptane or Octane at p= 0.1 MPa from (273.15 to 363.15) K. *Journal of Chemical and Engineering Data*, 52(4):1226–1232.

Lange, J.-P., van der Heide, E., van Buijtenen, J., and Price, R. (2012). Furfural - A Promising Platform for Lignocellulosic Biofuels. *ChemSusChem*, 5(1):150–166.

Lewis, G. N. (1901). Das Gesetz physiko-chemischer Vorgänge. *Zeitschrift für Physikalische Chemie*, 38(1):205–226.

Lladosa, E., Martínez, N. F., Montón, J. B., and de La Torre, J. (2011). Measurements and correlation of vapour–liquid equilibria of 2-butanone and hydrocarbons binary systems at two different pressures. *Fluid Phase Equilibria*, 307(1):24–29.

Lozano, L. M., Montero, E. A., Martín, M. C., and Villamañán, M. A. (1995). Vapor-liquid equilibria of binary mixtures containing methyl tert-butyl ether (MTBE) and/or substitution hydrocarbons at 298.15 K and 313.15 K. *Fluid Phase Equilibria*, 110(1-2):219–230.

Luther, S. K., Stehle, S., Weihs, K., Will, S., and Braeuer, A. (2015). Determination of Vapor-Liquid Equilibrium Data in Microfluidic Segmented Flows at Elevated Pressures using Raman Spectroscopy. *Analytical Chemistry*, 87(16):8165–8172.

Martin, M. L. and Youings, J. C. (1980). Vapour pressures and excess Gibbs free energies of cyclohexane + n-hexane, + n-heptane and + n-octane at 298.15 K. *Australian Journal of Chemistry*, 33(10):2133–2138.

Mohsen-Nia, M., Modarress, H., and Alimohammady, F. (2009). Densities and Viscosities of Binary Mixtures of Poly(vinyl chloride) and Tetrahydrofuran at Temperatures (283.15 to 303.15) K. *Journal of Chemical and Engineering Data*, 54(4):1375–1377.

Montgomery, C. J., Cremer, M. A., Chen, J.-Y., Berkeley, B., Westbrook, C. K., and Maurice, L. Q. (2002). Reduced Chemical Kinetic Mechanisms for Hydrocarbon Fuels. *Journal of Propulsion and Power*, 18(1):192–198.

Narasigadu, C. (2011). *Design of a static micro-cell for phase equilibrium measurements: measurements and modelling*. PhD Thesis, University of KwaZulu Natal and L'Ecole Nationale Supérieure des Mines de Paris, Durban.

Negadi, L., Mokbel, I., Negadi, A., Kaci, A. A., Jose, J., and Gmehling, J. (2009). Isothermal Vapor-Liquid Equilibria and Excess Enthalpy Data for the Binary System (Butyric Acid + Toluene). *Journal of Chemical and Engineering Data*, 54(7):2045–2048.

Pádua, A. A. H., Fareleira, J. M. N. A., Calado, J. C. G., and Wakeham, W. A. (1996). Density and Viscosity Measurements of 2,2,4-Trimethylpentane (Isooctane) from 198 K to 348 K and up to 100 MPa. *Journal of Chemical and Engineering Data*, 41(6):1488–1494.

Pelletier, M. J. (2003). Quantitative analysis using Raman spectrometry. *Applied Spectroscopy*, 57(1):20A–42A.

Peng, D.-Y. and Robinson, D. B. (1976). A New Two-Constant Equation of State. *Industrial & Engineering Chemistry Fundamentals*, 15(1):59–64.

Peper, S. and Dohrn, R. (2012). Sampling from fluid mixtures under high pressure: Review, case study and evaluation. *The Journal of Supercritical Fluids*, 66(0):2–15.

Peukert, S., Naumann, C., Braun-Unkhoff, M., and Riedel, U. (2011). Formation of H-atoms in the pyrolysis of cyclohexane and 1-hexene: A shock tube and modeling study. *International Journal of Chemical Kinetics*, 43(3):107–119.

Pfennig, A. (2004). *Thermodynamik der Gemische.* Springer Verlag Berlin Heidelberg.

Ra, Y. and Reitz, R. D. (2011). A combustion model for IC engine combustion simulations with multi-component fuels. *Combustion and Flame*, 158(1):69–90.

Raal, J. D. and Mühlbauer, A. L. (1998). *Phase equilibria: Measurement and computation.* Series in chemical and mechanical engineering. Taylor & Francis, Washington, DC.

Raal, J. D. and Ramjugernath, D. (2005). Vapour-Liquid Equilibrium at Low Pressure. In Weir, R. D. and de Loos, T. W., editors, *Measurement of the thermodynamic properties of multiple phases*, volume no. 41 of *IUPAC chemical data series*, pages 71–87. Gulf Professional Publishing and Elsevier, Amsterdam and London.

Redlich, O. and Kwong, J. N. S. (1949). On the Thermodynamics of Solutions. V. An Equation of State. Fugacities of Gaseous Solutions. *Chemical reviews*, 44(1):233–244.

Renon, H. and Prausnitz, J. M. (1968). Local compositions in thermodynamic excess functions for liquid mixtures. *AIChE Journal*, 14(1):135–144.

Rodrigues, W. L., Mattedi, S., and Abreu, J. C. N. (2005). Experimental vapor-liquid equilibria data for binary mixtures of xylene isomers. *Brazilian Journal of Chemical Engineering*, 22(3).

Saez, C., Compostizo, A., Rubio, R. G., Crespo Colin, A., and Díaz Peña, M. (1985). p, T, x, y data of benzene + n-hexane and cyclohexane + n-heptane systems. *Fluid Phase Equilibria*, 24(3):241–258.

Schrötter, H. W. and Klöckner, H. W. (1979). Raman Scattering Cross Sections in Gases and Liquids. In Weber, A., editor, *Raman Spectroscopy of Gases and Liquids*, volume 11 of *Topics in Current Physics*, pages 123–166. Springer, Berlin and Heidelberg.

Schuster, J. J., Will, S., Leipertz, A., and Braeuer, A. (2014). Deconvolution of Raman spectra for the quantification of ternary high-pressure phase equilibria composed of carbon dioxide, water and organic solvent. *Journal of Raman Spectroscopy*, 45(3):246–252.

Serrano-Ruiz, J. C., West, R. M., and Dumesic, J. A. (2010). Catalytic Conversion of Renewable Biomass Resources to Fuels and Chemicals. *Annual Review of Chemical and Biomolecular Engineering*, 1(1):79–100.

Sholl, D. S. and Lively, R. P. (2016). Seven chemical separations to change the world. *Nature*, 532(7600):435–437.

Sieg, L. (1950). Flüssigkeit-Dampf-Gleichgewichte in binären Systemen von Kohlenwasserstoffen verschiedenen Typs. *Chemie Ingenieur Technik*, 22(15):322–326.

Sims, R. E., Mabee, W., Saddler, J. N., and Taylor, M. (2010). An overview of second generation biofuel technologies. *Bioresource Technology*, 101(6):1570–1580.

Sirignano, W. A. (2014). *Fluid Dynamics and Transport of Droplets and Sprays*. Cambridge University Press, New York, NY, USA, 2nd edition edition.

Soave, G. (1972). Equilibrium constants from a modified Redlich-Kwong equation of state. *Chemical Engineering Science*, 27(6):1197–1203.

Span, R., Lemmon, E. W., Jacobsen, R. T., Wagner, W., and Yokozeki, A. (2000). A Reference Equation of State for the Thermodynamic Properties of Nitrogen for Temperatures from 63.151 to 1000 K and Pressures to 2200 MPa. *Journal of Physical and Chemical Reference Data*, 29(6):1361–1433.

Span, R. and Wagner, W. (1996). A New Equation of State for Carbon Dioxide Covering the Fluid Region from the Triple–Point Temperature to 1100 K at Pressures up to 800 MPa. *Journal of Physical and Chemical Reference Data*, 25(6):1509–1596.

Steinhauser, H. H. and White, R. R. (1949). Vapor-Liquid Equilibria Data for Ternary Mixtures: Methyl Ethyl Keton-n-Heptane-Toluene System. *Industrial and Engineering Chemistry*, 41(12):2912–2920.

Stratmann, A. and Schweiger, G. (2002). Fluid phase equilibria of ethanol and carbon dioxide mixtures with concentration measurements by Raman spectroscopy. *Applied Spectroscopy*, 56(6):783–788.

Takeo, M., Nishi, K., Nitta, T., and Katayama, T. (1979). Isothermal vapor-liquid equilibria for two binary mixtures of heptane with 2-butanone and 4-4-methyl-2-pentanone measured by a dynamic still with a pressure regulation. *Fluid Phase Equilibria*, 3(2-3):123–131.

Tanbonliong, J. O. and Prausnitz, J. M. (1997). Vapour-liquid equilibria for some binary and ternary polymer solutions. *Polymer*, 38(23):5775–5783.

Tang, X. and Gross, J. (2010). Renormalization-Group Corrections to the Perturbed-Chain Statistical Associating Fluid Theory for Binary Mixtures. *Industrial & Engineering Chemistry Research*, 49(19):9436–9444.

Thewes, M., Muether, M., Pischinger, S., Budde, M., Brunn, A., Sehr, A., Adomeit, P., and Klankermayer, J. (2011). Analysis of the Impact of 2-Methylfuran on Mixture Formation and Combustion in a Direct-Injection Spark-Ignition Engine. *Energy Fuels*, 25(12):5549–5561.

Valderrama, J. O. (2003). The State of the Cubic Equations of State. *Industrial & Engineering Chemistry Research*, 42(8):1603–1618.

Vallés, C., Pérez, E., Mainar, A. M., Santafé, J., and Domínguez, M. (2006). Excess Enthalpy, Density, Speed of Sound, and Viscosity for 2-Methyltetrahydrofuran + 1-Butanol at (283.15, 298.15, and, 313.15) K. *Journal of Chemical and Engineering Data*, 51(3):1105–1109.

van der Waals (1873). *Over de continuiteit van den gas- en vloeistoftoestand*. Dissertation, Leiden University, Leiden.

van Ness, H. C. and Abbott, M. M. (1978). A procedure for rapid degassing of liquids. *Industrial & Engineering Chemistry Fundamentals*, 17(1):66–67.

van Nhu, N. and Kohler, F. (1989). Excess properties of mixtures of polar components and hydrocarbons of varying local polarisability. *Fluid Phase Equilibria*, 50(3):267–296.

van Nhu, N., Singh, M., and Leonhard, K. (2008). Quantum Mechanically Based Estimation of Perturbed-Chain Polar Statistical Associating Fluid Theory Parameters for Analyzing Their Physical Significance and Predicting Properties. *The Journal of Physical Chemistry B*, 112(18):5693–5701.

Vittal Prasad, T. E., Sujana, V., Manasa, M., and Prasad, D. H. L. (2007). Vapor–liquid equilibria of some binary mixtures formed by methylethylketone at 94.7 kPa. *Physics and Chemistry of Liquids*, 45(4):419–423.

Wisniak, J., Embon, G., and Shafir, R. (1997). Isobaric Vapor−Liquid Equilibria in the Systems Methyl 1,1-Dimethylethyl Ether + Methyl Ethanoate and Oxolane + Heptane. *Journal of Chemical and Engineering Data*, 42(4):681–684.

Wisniak, J., Fishman, E., and Shaulitch, R. (1998). Isobaric Vapor−Liquid Equilibria in the Systems 2-Butanone + Heptane and 2-Butanone + Oxolane. *Journal of Chemical and Engineering Data*, 43(4):537–540.

Xue, Z., Mu, T., and Gmehling, J. (2012). Comparison of the a Priori COSMO-RS Models and Group Contribution Methods: Original UNIFAC, Modified UNIFAC(Do), and Modified UNIFAC(Do) Consortium. *Industrial & Engineering Chemistry Research*, 51(36):11809–11817.

Yan, X.-H., Wang, Q., Chen, G.-H., and Han, S.-J. (2001). Azeotropes at Elevated Pressures for Systems Involving Cyclohexane, 2-Propanol, Ethyl Acetate, and Butanone. *Journal of Chemical and Engineering Data*, 46(5):1235–1238.

Young, K. L., Mentzer, R. A., Greenkorn, R. A., and Chao, K. C. (1977). Vapor-liquid equilibrium in mixtures of cyclohexane + benzene, + octene-1, + m-xylene, and + n-heptane. *The Journal of Chemical Thermodynamics*, 9(10):979–985.

Zhang, Z., Jia, P., Huang, D., Lv, M., Du, Y., and Li, W. (2013). Vapor–Liquid Equilibrium for Ternary and Binary Mixtures of Tetrahydrofuran, Cyclohexane, and 1,2-Propanediol at 101.3 kPa. *Journal of Chemical and Engineering Data*, 58(11):3054–3060.